Rogers Locomotive and Machine Works

The Rogers Locomotive Company, Paterson, New Jersey.

Rogers Locomotive and Machine Works

The Rogers Locomotive Company, Paterson, New Jersey.

ISBN/EAN: 9783743377110

Manufactured in Europe, USA, Canada, Australia, Japa

Cover: Foto ©berggeist007 / pixelio.de

Manufactured and distributed by brebook publishing software
(www.brebook.com)

Rogers Locomotive and Machine Works

The Rogers Locomotive Company, Paterson, New Jersey.

Thomas Rogers

THE

ROGERS LOCOMOTIVE COMPANY,

PATERSON, - NEW JERSEY.

ESTABLISHED IN 1831.

NEW YORK
GEORGE G. PECK, PRINTER,
[illegible] CHAMBERS STREET.
1897.

ROGERS LOCOMOTIVE COMPANY.

The works of this company were established by Thomas Rogers in the year 1831, and from 1832 were carried on under the firm name of Rogers, Ketchum & Grosvenor until the death of Thomas Rogers in the year 1856, when they were reorganized under a charter with the title of The Rogers Locomotive & Machine Works, with his son, J. S. Rogers, as President, who from that time until 1893 continued in that capacity to manage the business of these works. From February, 1893, they were continued under the name of the Rogers Locomotive Company with Robert S. Hughes as President, and Reuben Wells as Superintendent.

Since Thomas Rogers commenced to build locomotives in Paterson not only has the machine as a whole been going through a process of evolution, but there has also been a development of its various parts to the functions which they have to perform.

The "Sandusky," the first locomotive built by Thomas Rogers in 1837, weighed probably less than ten tons. Since then the weights have gradually increased, and it is now no uncommon thing to build locomotives in these works weighing seventy-five tons and over. The working pressure in the boilers has also been increased from about one hundred pounds per square inch in the first Rogers engines built to one hundred and eighty, and in some cases to two hundred pounds, in engines built during the past few years.

Within the past five years a number of new and improved tools have been added, specially designed for the work to be done, so that the shops are now equipped with the most approved modern tools for doing accurate work, and with a complete system of templates and gauges by the use of which the similar parts of locomotives of the same class are furnished with a degree of precision which insures their being interchangeable, thus making it practicable to supply duplicate parts of locomotives built at these works at the shortest notice.

The Company has now every facility which long experience, thorough organization and abundant capital can provide for turning out high class work and for conducting the business of manufacturing locomotives in the most perfect manner possible.

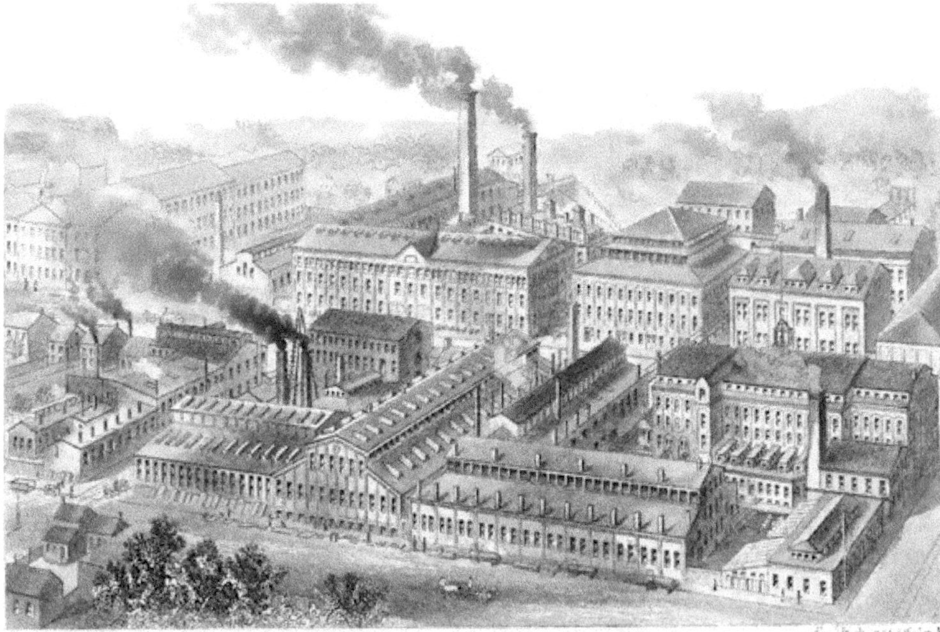

Rogers Locomotive Works

THE FOLLOWING IS A GENERAL DESCRIPTION BLANK USED IN GIVING THE DIMENSIONS, WEIGHTS AND OTHER DETAILS IN THE CONSTRUCTION AND EQUIPMENT OF WHAT ARE CALLED THE ROGERS STANDARD LOCOMOTIVES, AND A COPY OF SPECIFICATIONS FOR THE MATERIALS USED UNLESS A DIFFERENT ONE IS DESIRED BY THE PURCHASER.

GENERAL DESCRIPTION OF LOCOMOTIVE AND TENDER.

Gauge Cylinders pairs of Driving Wheels in. diam.
and Wheel Truck. Boiler top
No. Injector Wheel Tender gallon Tank
Fixed wheel base of Engine Total wheel base of Engine
Weight on Drivers about Total wheel base of Engine and Tender
Weight on Truck about Total weight of Engine in working order about
Total weight of Tender loaded

GENERAL DESCRIPTION.

General Design Illustrated by Photograph Below.

Fuel. Fuel

Boiler. top Boiler, with extended front, to be made of the best quality of properly braced, longitudinal seams to be butt jointed and quadruple or sextuple riveted wherever practicable with inside and outside welt pieces, with rivets of suitable size alternated on each side of main seam. Extra pieces riveted to inside of side sheets, providing double thickness for studs of expansion braces.

Steam Pressure. To sustain a working pressure of pounds per square inch. All plates planed at edges.

Boiler Diam. Outside diameter of Shell at Smoke Box inches.

Thickness of Boiler plates, cylindrical part inch, slope sheet inch, wagon-top inch, throat sheet inch, back head sheet inch, side sheets of shell inch. Cleaning plugs in corners, and in back head and on sides at top of crown sheet at suitable points for washing and in front flue sheet. Blow-off cock in front or rear of furnace.

Stay Bolts. Stay bolts to be of

The upper rows to be inch diameter, the other rows inch diameter and not over inches from centre to centre.

Crown Bars. Crown Bars to be made of two bars welded together at end and in number inches centre to centre, to have good bearing on sides of furnace and fitted with round taper washers down to the Crown Sheet, covering as little space as possible, and secured to Crown Sheet by button head bolts inch in diameter with nut and washer on top of crown bar. Crown bars securely braced to the shell of the boiler and dome with sling stays, flat surface on the sides of the connection between wagon top and cylinder part to be ribbed and braced across from side to side above flues.

Radial Stays. The Crown Sheet is supported by radial stays, the stays which are inch diameter in the body and inch diameter at the thread in the crown sheet, are screwed through crown and shell sheets and spaced inches between centres. The centre six or more longitudinal rows have button heads under the crown sheet.

Belpaire Firebox With the Belpaire type of Firebox the crown sheet is supported by vertical stays inch diameter in the body and inch diameter at the thread in the crown sheet, screwed through the crown and roof sheets, and spaced inches between centres. The flat sides above the crown sheets are braced by rows of cross stays at $1\frac{1}{8}$ inch diameter in the body and $1\frac{1}{4}$ inch diameter where screwed through the sheets and the lower two rows are nutted on the outside of the sheets.

The back head and front flue sheets of all boilers are thoroughly braced by heavy T irons and by longitudinal braces to the shell of the boiler. The Firebox Flue Sheet is also strongly braced to the barrel of the boiler below the flues.

Furnace. Furnace, inside length about inches; width about inches; height, front about inches; back inches. Furnace plates of inch thick, sides and back. Crown sheet inch thick
Furnace Flue Sheet inch thick.
Flue Sheet Smoke Box inch thick.

Test. Boiler tested to pounds pressure per square inch.

Flues. About to be of long in outside diameter, wire gauge and of best quality.

Dome. inches diameter, inches high.

Safety Valves. lock up valves fitted with relief lever.

Dry Pipe. Dry Pipe of lap welded iron inches outside diameter inch thick.

Throttle Valve. In dome, double seat balance poppet pattern.

Steam Pipes and Exhaust Pipes. Steam Pipes and Exhaust Pipes of cast iron of suitable area. Exhaust nozzle.

Smoke Stack. Smoke Stack for burning of approved design and proportions pattern.

Extended Smoke Box. To be inches long, fitted with air tight front and door, spark outlet, deflecting plate and steel wire netting all in the most approved manner.

Grate Bars. Grate Bars of cast iron with drop plate at front.

Ash Pan. Ash Pan with damper at each end operated from foot board.

Fire Door. The Fire Hole Ring and its door of cast iron with planed joint; the ring well fitted and held to the boiler head by studs.

Brick Arch. Supported on

Jacket. Boiler lagged with and covered with iron. Bands of

Frame. Frame to be of best hammered iron made in two sections. Pedestals forged solid with main frame and protected from wear of boxes by cast iron gibs and wedges. Front rails bolted and keyed to main frames and with front and back lugs forged on for cylinder connections. Frames inches wide, planed and fitted to standard dimensions. The Pedestals to have one movable and one fixed gib. Pedestal Caps secured by two well fitted bolts on each end.

Foot Plate. Foot Plate securely bolted and lugged to frame.

Cylinders. Cylinders inches diameter, inches stroke, each cylinder and half saddle cast in one piece, reversible and interchangeable, to be made of close grained cast iron as hard as can be worked, and accurately fitted and drilled to templates. Valve seat raised inches above the steam chest seat.

Driving Wheels. pairs of Driving Wheels inches diameter outside of Tires. Tires of steel inch thick when finished flanged plain.

Wheel Centres of spokes, turned to inches diameter, correctly quartered.

Counterbalance. Each wheel counterbalanced accurately to the weight of its crank pin and hub, and of the rods resting on its pin, and also to an equal portion of the total per cent. of the reciprocating weight that is to be balanced. The reciprocating parts are all made as light as is consistent with the necessary strength and wearing surfaces.

Engine Truck. Engine Truck to be centre bearing and to have Wheels inches diameter. Axles of the best quality of inches diameter with Journal inches.

Driving Axles. Driving Axles to be of best quality of inches diameter, journals inches diameter, inches long.

Crank Pins. Crank Pins of and of good proportions.

Driving Boxes. Driving Boxes of best quality of with round top brasses pressed in from the side, and pinned with ⅞ inch diameter brass pins at the top quarters of the box.

All driving box and truck box bearings to be of the best

Connecting Rods. Connecting Rods to be of well forged Body section. Front end fitted with and brasses. Back end fitted with

Oil cups on

Parallel Rods. Parallel Rods to be of well forged Body section. At pins, fitted with , at Pins with Oil cups on.

All rod bearings to be of the best quality of

Piston Rods. Piston Rods to be of inches in diameter, fitted with packing.

Pistons. Piston and Followers to be of cast Packing rings, cast iron of approved pattern.

Valve Stems. Fitted with packing.

Cross Heads. Cross Heads of with bearings of

Pattern of crosshead of type.

Guides. Guides to be of of the type finished and well secured to cylinder head and to the guide yoke.

Valve Motion. Valve Motion to be Shifting link motion with yoke at driving axle and straight eccentric rods, graduated to cut off equally at all points of the stroke and thoroughly case hardened. Links inches wide on the face. Sliding blocks with long flanges.

Slide Valve. Slide Valve of cast iron balanced. Valves oiled from Sight feed oilers in cab through copper pipes under jacket.

Rockers. Rocker Shafts of the best hammered iron. Rocker Boxes of cast iron.

Eccentrics and Straps. Eccentrics and straps of cast iron with face inches wide. Large oil chambers on both upper and lower side.

Eccentrics secured to driving axle by set screws and keys.

Lifting Shaft. Lifting shaft of wrought iron, with solid arms forged on.

Pumps and Injectors. Pump Injectors.

Equalizers. Equalizers of best hammered iron, fitted with steel bearings for hanger gibs.

Springs. Springs of approved make of cast steel. Tempered in oil
Spring hangers of refined iron, fitted with steel gibs.

Expansion Plates Expansion plates of flange steel of suitable dimensions, well secured by studs and bolts.

Pilot. Pilot of hard wood, iron bound, substantially built and well braced.

Draw Head. Draw head of type.

Bumper. Bumper of hard wood, extending to outside of each cylinder, securely fastened to Engine Frame, covered and faced with Iron and backed by steel plate ½ inch thick, the same width as bumper and about 7 feet long.

Cab. Cab to be of painted, double thick, glass window, tin roof, fitted with iron corner and bracket pieces at running board and at roof, fastened to a cast iron "arch" piece attached to the top of the boiler, and all securely bolted together. Back end to be type.

Doors and sash provided with substantial fastenings. Cab fitted with signal gong, cab lamp, cushioned seats, arm rests and the necessary tool boxes.

Running Boards. Running boards of with T iron edge polished.

Hand Rails. Hand rails of polished. Columns of iron polished.

Oil Cups. Oil cups of the proper size are provided at all places where necessary.

Sundries. Steam gauge, sand box, head light stand, bell, whistle, blower, gauge cocks, and turret with dry pipe to dome inside the boiler, fitted with a shut off valve for supplying steam to injectors, air brake and blower.

TENDER.

Frame. Tender frame to be of made and braced in the most substantial manner. To have two four wheel Trucks of

Truck. pattern. Each truck to have two pairs of wheels inches diameter. Upper bolsters of

Springs. lower bolsters of Springs pattern.

Tank. Axles to be of best with journals inches. Tank

Axles. of Steel, of gallons (231 cubic inches) capacity, with space for tons of coal.
Top, inside and bottom plates thick; outside plates thick.

Brake. Tender fitted with a powerful hand brake, an iron bound wooden tool box on back end of frame and two on top of the tank.

Drawhead. At back of tender

Tools. Engine and Tender to have a complete set of all the usual tools, such as hammers, chisels, wrenches, pinch bar, jackscrews, fire tools and oil cans all of best quality.

General. All material and workmanship to be of the best quality, work accurately fitted, all parts to bear perfect proportion to each other, rendering the whole machine, when complete, first-class in every respect. Each engine will be run by steam before leaving the builders' shops. All principal parts accurately fitted to gauges, and thoroughly interchangeable. All bolts and nuts of United States Standard thread. Engine and Tender to be well painted and varnished. Lettering and numbering to be as specified by the purchaser

Brake.

Train Signal.

Steam Heat.

Head Lamp.

SPECIFICATIONS FOR MATERIALS.

The materials used for the construction of this locomotive are in conformity with the following requirements, unless the use of special material or specifications are requested by the purchaser.

Boiler and Fire-box Steel. Flange Steel, tensile strength 55,000 to 65,000 lbs. per square inch of section, elongation not less than 20 per cent. in 8 inches.

Test piece, after having rough edges removed, must be able, without annealing, to bend over on itself both hot and cold, and after being heated to a cherry red and dipped in water at 80 degrees, without showing cracks or flaws on outside edge.

Firebox Steel, tensile strength 50,000 to 58,000 lbs. per square inch of section, elongation not less than 22 per cent. in 8 inches.

Plates will be rejected if analysis shows them to contain—Carbon, over .25 or below .15; Phosphorus, over .035; Manganese, over .45; Sulphur, over .035; Silicon, over .03.

Should any plates develop defects in working, they will be rejected.

Staybolt Iron. Tensile strength should be 49,000 lbs. with an elongation of 28 per cent. in 8 inches.

If the tensile strength is less than 47,000 lbs., or the elongation less than 26 per cent., or if the fracture is crystalline, the iron will be rejected. Iron must be free from seams, and take a good thread with dies in fair working order. Steel will not be accepted.

Boiler Tubes. All tubes must be lap welded and made of charcoal iron. A careful examination will be made of each tube, and those showing imperfect welds, pit holes, or other defects, will be rejected.

Tubes must be straight and not over 1/32 inch out of round. The thickness at any part must not be less than the gauge ordered, nor greater at the thickest part than one gauge more than ordered.

Tubes must stand the following Test:

A section 1 1/4 inch long, taken at random, must stand hammering down vertically until solid without cracking or splitting. They must stand expanding and bending over without crack or flaw.

Tank Steel. Sheets to be of uniform thickness, sheared square and closely to sizes ordered, with a variation of not over 7 per cent. from the estimated weight. The metal must be soft, homogeneous steel, free from surface defects. Sheets with rough irregular surface, pit holes, cinder spots, etc., will be rejected.

Test strips must stand bending hot or cold through 180 degrees over a mandrel whose diameter is 1 1/2 times the thickness of the sheet, without showing cracks or signs of fracture.

Steel Castings. Castings must be free from blow holes and shrinkage cracks.

Tensile strength should be 64,000 lbs., with an elongation of 16 per cent. in two inches. Castings having a tensile strength of less that 55,000 lbs., and an elongation of less than 10 per cent. in two inches, will be rejected. Castings badly warped or distorted, which will not true up properly to drawing, will be rejected.

Bar Iron. Must be thoroughly welded, free from seams, blisters and cinder spots, with a fibrous fracture, free from crystalization. A greater variation than 2 1/2 per cent. from the estimated weight will not be allowed.

The tensile strength should be 50,000 lbs. per square inch, and an elongation of 18 per cent. in 8 inches.

Iron showing a tensile strength of less than 48,000 lbs., with an elongation of less than 14 per cent. in 8 inches, will be rejected.

Steel Forgings. The steel blooms used for the forgings of axles, crank pins, main and parallel rods, piston rods, etc., must be of open hearth steel having a tensile strength of 85,000 lbs. per square inch, with an elongation of 20 per cent. in two inches.

Material with a tensile strength of less than 80,000 lbs. or an elongation of less than 15 per cent., or with phosphorus exceeding .05, will be rejected.

Steel Shapes. Angles channels, tees, etc., must be free from injurious seams, &c., variation from estimated weight not to exceed 5 per cent.

Tensile strength from 50,000 to 67,000 lbs., with an elongation of not less than 15 per cent. in 8 inches. Specimens must stand bending through 180 degrees to an inner diameter equal to its own thickness, without crack or flaw.

Cast Iron Wheels. Will be furnished with guaranteed mileage as follows :—

26 in. and 28 in.	wheels	not	less	than	40,000	miles.
30 in.	"	"	"	"	45,000	"
33 in.	"	"	"	"	50,000	"

According to Master Car Builders' Association recommended practice, adopted 1893.

Or if preferred wheels will be furnished from approved specifications or make, with drop test, but without guarantee.

Firebox Copper. Plates for fireboxes should contain not less than 99.5 per cent. of pure copper, and be free from flaws, cracks or other defects.

Tensile strength must not be less than 30,000 lbs. per square inch, with an elongation of not less than 20 per cent. in 2 inches.

Copper Stay Bolts. Copper for stay bolts should contain not less than 99.5 per cent. of pure copper, and be free from defects.

The tensile strength must not be less than 30,000 lbs. per square inch, with an elongation of not less than 20 per cent. in two inches.

THE TRACTIVE POWER OF LOCOMOTIVES.

There is some difference in the figures given by various authorities to indicate the proportion which the friction or adhesion of the wheels on the rails bears to the weight on them. The figures which are perhaps used most in practice are those published in Molesworth's "Pocket-Book of Engineering Formulæ." These are as follows:

ADHESION PER TON OF 2,240 LBS. ON THE DRIVING-WHEELS.

When the rails are very dry	600 lbs. per ton.
When the rails are very wet	550 " " "
In ordinary English weather	450 " " "
In misty weather, if the rails are greasy	300 " " "
In frosty or snowy weather	200 " " "

In D. K. Clark's "Manual for Mechanical Engineers," page 724, he gives a report of experiments made by M. Poirée on the Paris & Lyons Railroad with a wagon by skidding the wheels. Of these experiments Clark says:

"At speeds under 20 miles per hour it appears from the table that, when the rails are dry, the co-efficient of friction, or the adhesion, is *one-fifth* of the weight, and that on very dry rails it is one-fourth. As the speed is increased, the adhesion is reduced. These data are corroborative of the results of the author's experiments on the ultimate tractive force of locomotives on dry rails, from which he obtained a co-efficient of friction equal to *one-fifth* of the weight, at speeds of about 10 miles per hour."

In the paper "On Effect of Brakes upon Railway Trains," read by Captain Galton before the Institution of Mechanicals Engineers,* the following determination of the adhesion of wheels is given. It must be kept in mind, too, that he makes the distinction between "adhesive" and sliding friction. By "adhesive" is meant the friction between rolling wheels and the track:

"On dry rails it was found that the co-efficient of adhesion of the wheels was generally over 0.20. In some cases it rose to 0.25 or even higher. On wet or greasy rails without sand, it fell as low as 0.15 in an experiment, but averaged about 0.18. With the use of sand on wet rails it was above 0.20 at all times; and when the sand was applied at the moment of a starting, so that the wind of the rotating wheels did not blow it away, it rose up to 0.35, and even above 0.40."

This is probably the most correct determination of the adhesion of wheels that has so far been made, and shows that the ordinary rule of taking the adhesion at *one-fifth* of the weight in the driving-wheels is quite within the limits of ordinary practice. Even on a wet or greasy rail, with the use of sand, it was above 0.20 at all times. Under favorable conditions, without sand, it may in making calculations be estimated to be *one-fourth*.

* See *Engineering*, of May 2, 1879, page 371.

In order to put these figures in a form in which they can easily be remembered and conveniently used, they may be given as follows:

ADHESION OF LOCOMOTIVES.

Under ordinary conditions, without using sand on the rails, or on wet sanded rails	One-fifth the weight on the driving-wheels.
Under favorable conditions without sand	One-fourth the weight on the driving-wheels.
On a dry sanded rail	One-third the weight on the driving-wheels.

These may be taken as working data. It may be stated that the most recent experiments have shown that the resistance of good American cars does not exceed 6 lbs. per ton of 2,000 lbs. at slow speeds on a straight and level track, and when in the best condition and good weather it is probably not over 4 lbs. The wind, however, has an important influence, and as this is very variable it is hardly safe to take the resistance, under the conditions named above, at less than 6 lbs. per ton.

With reference to the influence of speed on the resistance, it must be admitted that our knowledge is somewhat inexact, and probably the law or laws which govern it are not fully understood. The following rule, though, will give results which do not differ materially from those given by the most reliable experiments which have thus far been made.

To get the resistance per ton (of 2,000 lbs.) of a train on a straight and level track at any given speed:

Square the speed in miles per hour and divide by 171 and add 6.

To get the resistance per ton due to any grade:

Multiply the rise in feet per mile by 0.3788 and add the result to the resistance due to the speed on a straight and level track.

Our knowledge of the resistance due to curves, like that due to speed, is not very exact, but the most reliable information we have indicates that the resistance is equal to about half a pound per ton per degree of curvature for a standard gauge track.

We may then tabulate these calculations as follows:

RESISTANCE OF TRAINS.

On straight and level track at low speeds	6 lbs. per ton of 2,000 lbs.
For resistance due to speed: *Square the speed in miles per hour and divide by 171*	"
For resistance due to grade: *Multiply the rise in feet per mile by 0.3788*	— "
For resistance due to curves: *Add ½ lb. per degree of curvature*	— "
Total	— lbs.

If the radius of the curve is given, the "degree" may be found approximately *by dividing the radius into* 5730. This rule is correct enough for ordinary curves of over 500 feet radius.

To calculate how much a locomotive will pull, the rule used by M. N. Forney in his Catechism of the Locomotive will be found to be as simple as any other, as follows:

"By multiplying together the area of the piston in square inches, the average steam pressure in pounds per square inch on the piston during the whole stroke, and four times the length of the stroke of the piston in inches and dividing the product by the circumference of the driving wheel in inches, will give the tractive power exerted in pounds."

To make the calculation it is, therefore, necessary to know or determine:

1st.—The diameter of the cylinder and length of stroke.

2d.—The diameter of driving wheels.

3d.—The average steam pressure on the piston for the whole length of the stroke.

4th.—Grade and degrees of curvature of the road.

5th.—Speed at which the train is to be hauled.

For example, suppose we want to calculate how much a consolidation locomotive with 20 inch cylinders, a stroke of 26 inches and drivers 54 inches diameter will pull up a grade of 70 feet per mile with 9° curves and at a speed of 15 miles per hour, assuming that the average pressure on the pistons for the whole length of the stroke is 125 lbs. per square inch.

The cylinders being 20 inches in diameter the area of the piston would be 314.16 square inches: the average pressure on it 314.16 × 125 = 39,270 pounds. As each piston moves through 52 inches during one revolution, the total for the two (four times the length of the stroke) would be 104 inches. Therefore, 39,270 × 104 = 4,084,080, and as the circumference of the driving wheel is 169.6 inches, the tractive force of such an engine would be 4,084,080 ÷ 169.6 = 24,080 lbs. As the weight necessary to give the requisite amount of adhesion should be about five (5) times the tractive force, developed when doing full work, this locomotive should have 24,080 × 5 = 120,400 lbs. on its drivers, and its total weight would be about 138,000 lbs. The weight of the tender would be, say about 74,000 lbs., making a total of 106 tons for engine and tender.

The resistance per ton would be as follows:

Resistance on straight and level track		6.0 lbs.
" due to speed	$= \frac{15 \times 15}{171} =$	1.3 "
" grade	$= 70 \times .3788 =$	26.5 "
" curve	$= 9 \times \frac{1}{2}$	4.5 "
Total		38.3 lbs.

Therefore, as each ton will have a resistance of 38.3 lbs., and as the engine is capable of exerting a tractive force of 24,080 lbs. the total load which it can pull up such a grade would be represented by $\frac{24,080}{38.3} = 628.7$ tons. As the engine and tender weigh about 106 tons the train which the engine would pull will be represented by 628.7—106 = 522.7 tons. Such a locomotive at a train speed of 15 miles per hour would have a piston speed of about 400 feet per minute. To make up the loss of steam pressure from friction in passing through the pipes and steam ports and from compression at that piston speed so as to give an average effective pressure on the pistons of 125 lbs. per square inch for the whole length of the stroke, would require a boiler pressure of about 170 lbs. Indicator tests show that as the speed of the piston increases, the average effective steam pressure on it decreases, other things being equal, due, of course, to the causes mentioned above.

DIFFERENCE BETWEEN BOILER PRESSURE AND THE AVERAGE EFFECTIVE ON THE PISTONS FOR THE WHOLE LENGTH OF THE STROKE.

The following data given in table No. 1, and as illustrated in the diagram Fig. 2, was obtained from the *averages* of a large number of the most reliable indicator tests available made with locomotives built by nearly all the different builders in this country, and also locomotives built by a number of Railroad Companies in their own shops. In many cases the point of cut-off at the time was not known, but as the tests were made mostly to determine the power the engine would develop, it is most likely the cut-off in such cases was at the point where the engine would do its maximum work at that piston speed. It is probable that the averages in some cases were reduced slightly by more or less wire drawing at the throttle valve, or from the valve motion not being of the best in its proportions. It is therefore likely that with a good valve motion large ports and steam passages, and a wide open throttle, the average effective pressure for the stroke will be found to be somewhat above rather than below the per cent. of boiler pressure given in the table and shown in the diagram, at all, except the slowest speeds. At any rate the figures as given are probably as reliable as any that can be obtained at present on the subject.

TABLE I.

Diameter of driving wheels in inches.	Approximate train speed in miles per hour.	Approximate piston speed, feet per minute.	Per cent. of the boiler pressure effective on the pistons, average for the whole length of the stroke.
50 to 54 inches.......	5 miles.	130 feet.	90 per cent.
50 " 54 " ...	10 "	260 "	83 " "
54 " 56 "	15 "	367 "	76 " "
54 " 56 "	20 "	488 "	66 " "
60 " 62 "	25 "	550 "	61 " "
60 " 62 "	30 "	660 "	52 " "
60 " 62 "	35 "	770 "	44 " "
66 " 68 "	40 "	800 "	42 " "
66 " 68 "	45 "	900 "	37 " "
68 " 72 "	50 "	960 "	35 " "
68 " 72 "	55 "	1050 "	33 " "
78 inches.............	60 "	1035 "	33 " "
78 "	65 "	1120 "	31 " "
84 "	70 "	1120 "	31 " "
84 "	75 "	1200 "	30 " "
90 "	80 "	1196 "	30 " "
96 "	80 "	1120 "	31 " "

TABLE 2.

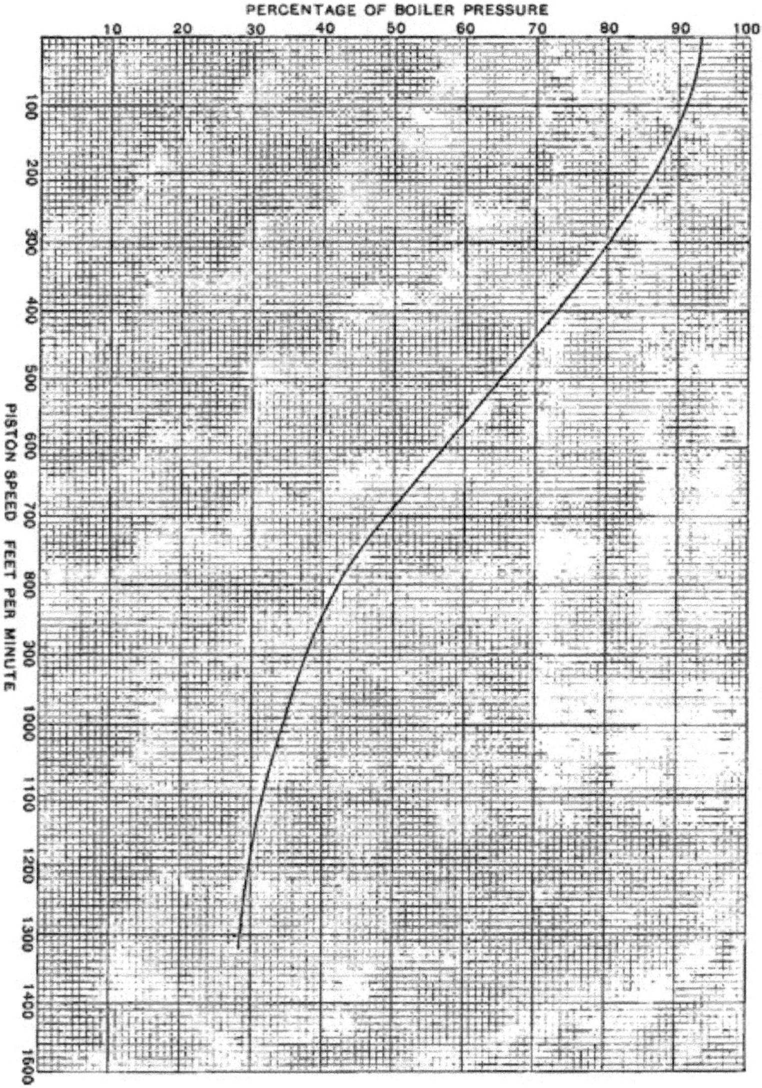
PERCENTAGE OF BOILER PRESSURE
10
20
30
40
50
60
70
80
90
100
100
200
300
400
500
600
700
800
900
1000
1100
1200
1300
1400
1500
PISTON SPEED FEET PER MINUTE

TABLE 3.

PISTON SPEED. FEET PER MINUTE.

24 Inch Stroke.

MILES PER HOUR.

Diam. of Drivers	Rev. per Mile	5 Rev.	5 Pist'n Speed	10 Rev.	10 Pist'n Speed	15 Rev.	15 Pist'n Speed	20 Rev.	20 Pist'n Speed	25 Rev.	25 Pist'n Speed	30 Rev.	30 Pist'n Speed	35 Rev.	35 Pist'n Speed	40 Rev.	40 Pist'n Speed
36	560.2	47	187	93	373	140	560	180	746	233	933	280	1120	326	1307	373	1493
37	545.2	45	182	91	363	136	545	181	726	227	908	272	1090	317	1272	363	1453
38	530.6	44	177	88	354	132	531	176	707	221	884	266	1061	309	1237	354	1414
39	517.2	43	172	86	345	129	517	172	689	215	862	258	1034	301	1207	345	1379
40	504	42	168	84	336	126	504	168	672	210	840	252	1008	294	1176	336	1344
41	491.9	41	164	82	327	123	491	163	654	204	818	245	982	286	1146	327	1309
42	480.3	40	160	80	320	120	480	160	640	200	800	240	960	280	1120	320	1280
43	469.0	39	156	78	313	117	469	156	625	195	781	234	937	273	1094	313	1250
44	458.4	38	153	76	306	114	458	152	611	191	764	229	917	267	1069	306	1222
45	448	37	149	75	298	112	448	149	597	186	746	224	896	261	1045	298	1194
46	438.4	36	146	73	292	109	438	146	584	182	730	219	876	255	1022	292	1168
47	429.1	35.7	143	71	286	107	429	143	572	178	714	214	858	250	1000	286	1143
48	420.1	35	140	70	280	105	420	140	560	175	700	210	840	245	980	280	1120
49	411.6	34	137	69	274	103	412	137	548	171	686	206	823	240	960	274	1097
50	403.3	33.6	134	67	269	101	403	134	537	168	672	202	806	235	941	269	1075
51	395.5	32.9	132	66	264	99	395	132	527	165	660	197	791	230	922	263	1054
52	387.7	32.3	129	65	258	97	388	129	516	161	646	194	775	226	904	258	1033
53	380.5	31.7	127	63	254	95	380	127	507	158	634	190	761	222	887	254	1014
54	373.4	31.1	124	62	249	93	373	124	497	155	622	187	747	217	870	249	996
55	366.6	30.5	122	61	244	92	367	122	488	152	611	183	732	213	855	244	977
56	360.2	30	120	60	240	90	360	120	480	150	600	180	720	210	840	240	960
57	353.8	29.5	118	59	236	88	354	117	471	147	589	177	707	206	825	236	943
58	347.7	29.0	116	58	232	87	348	116	463	145	579	174	695	202	811	232	927
59	341.7	28.4	114	57	228	85	342	113	455	142	569	171	683	199	797	228	911
60	336.1	28	112	56	224	84	336	112	448	140	560	168	672	196	784	224	896
61	330.6	27.5	110	55	220	82	331	110	440	138	551	165	661	193	771	220	881
62	325.3	27.1	108	54	217	81	325	108	433	135	542	163	650	190	758	217	867
63	320.1	26.6	107	53	213	80	320	106	426	133	533	160	640	186	747	213	853
64	315	26.2	105	52	210	79	315	105	420	131	525	157	630	183	735	210	840
65	310.2	25.8	103	51.7	207	77	310	103	413	129	517	155	620	181	724	207	827
66	305.6	25.4	102	51	204	76	306	101	407	127	509	153	611	178	713	204	814
67	301.0	25	100	50	201	75	301	100	401	125	501	150	602	175	702	201	802
68	296.6	24.7	99	49	198	74	297	98	395	123	494	148	593	173	692	198	791
69	292.2	24.3	97	48.7	195	73	292	97	390	122	487	146	584	170	682	195	779
70	288.1	24	96	48	192	72	288	96	384	120	480	144	576	168	672	192	768
71	284.0	23.6	95	47	189	71	284	94	378	118	473	142	567	166	662	189	757
72	280.1	23.3	93	46.6	187	70	280	93	373	116	466	140	560	163	653	187	747
73	276.3	23	92	46	184	69	276	92	368	115	460	138	552	161	644	184	737
74	272.5	22.7	90.8	45	182	68	272	90	363	113	454	136	545	159	636	182	726
75	268.9	22.4	89.6	44.8	179	67	269	89	358	112	448	134	537	157	627	179	717
76	265.3	22.1	88.4	44	177	66	265	88	354	110	442	132	530	154	619	177	707
78	258.6	21.5	86	43	172	64	258	86	345	108	431	129	517	151	603	172	689
80	252.1	21	84	42	168	63	252	84	336	105	420	126	504	147	588	168	672
84	240.1	20	80	40	160	60	240	80	320	100	400	120	480	140	560	160	640
86	234.5	19.5	78	39	156	58	234	78	312	97	390	117	469	136	547	156	625
90	224.1	18.7	75	37	149	56	224	74	299	93	373	112	448	130	523	149	598
96	210	17.5	70	35	140	53	210	70	280	87	350	105	420	122	490	140	560

TABLE 4.

PISTON SPEED. FEET PER MINUTE.

24 Inch Stroke.

MILES PER HOUR

Diam. of Drivers	Rev. per Mile	45 Rev.	45 Pist'n Speed	50 Rev.	50 Pist'n Speed	55 Rev.	55 Pist'n Speed	60 Rev.	60 Pist'n Speed	65 Rev.	65 Pist'n Speed	70 Rev.	70 Pist'n Speed	75 Rev.	75 Pist'n Speed	80 Rev.	80 Pist'n Speed
36	560.2																
37	545.2																
38	530.6	398	1592														
39	517.2	387	1552														
40	504.0	378	1512														
41	491.1	368	1473														
42	480.3	360	1441														
43	468.9	351	1407	390	1563												
44	458.4	342	1375	382	1528												
45	448.0	336	1344	373	1493												
46	438.4	327	1315	365	1461												
47	428.9	321	1287	357	1429	393	1572										
48	420.1	315	1260	350	1400	385	1540										
49	411.6	309	1236	343	1372	377	1509										
50	403.3	303	1209	336	1344	370	1478										
51	395.5	297	1185	329	1318	362	1450	395	1582								
52	387.7	291	1163	323	1292	355	1421	388	1551								
53	380.5	285	1141	317	1268	349	1395	380	1522								
54	373.5	280	1120	311	1245	342	1370	373	1494								
55	366.6	275	1100	305	1222	336	1344	367	1466	397	1589						
56	360.2	270	1080	300	1200	330	1320	360	1441	390	1561						
57	353.8	265	1061	295	1179	324	1297	354	1415	383	1533						
58	347.7	261	1043	290	1159	319	1275	348	1391	377	1507						
59	341.7	256	1026	285	1139	313	1253	342	1367	370	1481	399	1595				
60	336.1	252	1008	280	1120	308	1232	336	1344	364	1456	392	1568				
61	330.6	248	992	275	1102	303	1212	331	1322	358	1432	386	1543				
62	325.3	244	975	271	1084	298	1192	325	1301	352	1409	379	1518				
63	320.1	240	960	267	1067	293	1174	320	1280	347	1387	373	1494				
64	315.0	236	945	262	1050	289	1155	315	1260	341	1365	367	1470	393	1575		
65	310.2	233	931	258	1034	284	1137	310	1241	336	1344	362	1448	388	1551		
66	305.6	229	917	255	1019	280	1120	306	1222	331	1324	357	1426	382	1528		
67	301.0	225	903	251	1003	276	1103	301	1204	326	1304	351	1404	376	1504		
68	296.6	222	890	247	988	272	1087	297	1186	321	1285	346	1384	371	1483	395	1582
69	292.2	219	876	243	974	267	1071	292	1168	316	1265	341	1364	365	1461	390	1558
70	288.1	216	864	240	960	264	1056	288	1152	312	1248	336	1344	360	1440	384	1536
71	283.9	213	852	237	946	260	1041	284	1134	308	1230	331	1325	355	1419	378	1514
72	280.1	210	840	233	933	257	1027	280	1120	303	1213	327	1307	350	1400	373	1494
73	276.3	207	829	230	921	253	1013	276	1105	299	1197	322	1289	346	1381	368	1474
74	272.5	204	816	227	907	250	999	272	1090	295	1181	319	1271	342	1362	363	1452
75	268.9	201	807	224	896	246	986	269	1074	291	1164	314	1254	336	1344	358	1434
76	265.3	198	796	221	884	242	972	265	1060	287	1148	310	1238	330	1326	353	1414
78	258.6	194	776	215	862	237	948	259	1034	281	1120	302	1207	323	1293	345	1379
80	252.1	189	756	210	840	231	924	252	1008	273	1092	294	1176	315	1260	336	1344
84	240.1	180	720	200	800	220	880	240	960	260	1040	280	1120	300	1200	320	1280
86	234.5	175	703	195	781	215	860	234	938	254	1016	273	1094	293	1172	312	1250
90	224.1	168	672	187	747	205	822	224	896	243	971	261	1046	280	1120	299	1195
96	210.1	157	630	175	700	193	770	210	840	228	910	245	981	263	1050	280	1120

The piston speed in feet per minute for a stroke of 24 inches for the different speeds in miles per hour and drivers from 30 inches to 96 inches diameter is given in tables 3 and 4, pages 18 and 19. The two columns on the left give the diameter of the driver and the number of revolutions per mile due to its diameter.

To ascertain the piston speed per minute when the drivers are say 62 inches diameter and speed of engine 40 miles per hour take the figures in the vertical column for that speed (40 miles) corresponding to that giving the diameter and revolutions of the drivers, which in this case gives for the revolutions 272 and piston speed 867 feet per minute, due to a speed of 40 miles per hour. For a speed of 20 miles with 50 inch drivers the revolutions per minute as given are 134 and the piston speed 537 feet per minute, and so on.

TRACTIVE FORCE PER POUND OF EFFECTIVE PRESSURE ON THE PISTONS.

Tables 5 and 6, pages 22 and 23 give the tractive force at the rail for each pound of average effective pressure on the pistons for the whole length of the stroke. The figures in the horizontal column at top represents the diameter of drivers and the figures in the vertical column under that for the drivers, corresponding to that in the column of cylinder dimensions, gives the tractive force for each pound effective on the pistons. For instance, to find that due to 19 x 26 inch cylinders and drivers 62 inches diameter, take the figures in the column for a 62 inch diameter driver that are opposite the dimensions of the cylinders, which in this case is 151.2, that is, the engine will exert a tractive force at the rails of 151.2 pounds for each pound of average effective pressure on the pistons for the whole length of the stroke. In the case of an engine with 10 x 18 inch cylinders and 45 inch drivers the ratio would be 40 pounds of tractive force to each pound effective on the pistons, and so on.

TABLE 5.

CYLINDER. Dia.	Stroke.	26	27	28	29	30	31	32	33	34	35	36	37	38	39	40	41	42	43	44
		DIAMETER OF DRIVERS.																		
8	12	30.0	28.0	27.5	26.5	26.0	25.0	24.0	23.0	22.8	21.8	21.3	20.7	20.2	19.7	19.2	18.7	18.2	17.8	17.4
9	12	37.0	36.0	34.7	33.5	32.0	31.4	30.4	29.5	28.6	27.7	27.0	26.2	25.6	25.0	24.3	23.7	23.1	22.6	22.1
9	16	49.8	48.0	46.3	44.7	43.0	41.8	40.5	39.3	38.1	37.0	36.0	35.0	34.1	33.2	32.4	31.6	30.8	30.1	29.4
10	18	69.2	66.6	64.2	62.0	60.0	58.3	56.2	54.5	52.9	51.4	50.0	48.6	47.3	46.1	45.0	43.9	42.8	41.8	40.9
11	16	74.5	71.7	69.2	66.8	64.5	62.4	60.5	58.6	57.0	55.3	53.8	52.3	51.0	49.6	48.4	47.2	46.1	45.0	44.0
11	18	83.7	80.6	77.8	75.0	72.6	70.2	68.0	66.0	64.0	62.2	60.5	58.8	57.3	55.8	54.4	53.0	51.8	50.6	49.5
12	16	88.7	85.3	82.2	79.5	76.8	74.3	72.0	69.8	67.7	65.8	64.0	62.2	60.5	59.0	57.6	56.1	54.8	53.5	52.3
12	18	99.7	96.0	92.5	89.3	86.4	83.6	81.0	78.5	76.2	74.0	71.9	70.0	68.1	66.4	64.6	63.1	61.6	60.2	58.8
12	20	110.8	106.6	102.8	99.3	96.0	93.0	90.0	87.3	84.7	82.3	80.0	77.8	75.8	73.8	72.0	70.2	68.5	67.0	65.5
12	22	121.8	117.3	113.0	109.0	105.6	102.0	99.0	96.0	93.2	90.6	88.0	85.6	83.5	81.2	79.2	77.3	75.5	73.6	72.0
13	18	117.0	112.6	108.5	104.9	101.4	98.1	95.0	92.1	89.5	87.0	84.5	82.3	80.0	78.0	76.0	74.3	72.5	70.7	69.1
13	20	....	....	....	116.5	112.6	109.0	105.5	102.4	99.5	96.6	94.0	91.4	89.0	86.6	84.5	82.5	80.5	78.6	76.8
13	22	....	....	....	....	124.0	119.9	116.1	112.7	109.3	106.1	103.2	100.5	97.9	95.5	93.0	90.6	88.5	86.4	84.5
14	18	....	....	....	....	....	....	....	106.9	103.8	100.8	98.0	95.5	93.0	90.5	88.2	86.0	84.0	82.1	80.2
14	20	....						....	....	115.2	112.0	108.9	105.9	104.0	100.4	98.0	95.6	93.4	91.2	89.0
14	22	....						....	....	....	123.2	119.7	116.5	113.4	110.5	107.8	105.0	102.7	100.1	98.0
14	24							....	....	....	....	....	127.1	123.8	120.6	117.6	114.8	112.0	109.3	106.9
15	18							....	....	....	....	....	109.0	106.5	103.9	101.2	98.9	96.5	94.4	92.2
15	20							....	....	....	....	....	121.6	118.4	115.3	112.5	109.7	107.0	104.6	102.0
15	22								....	....	....	....	....	130.2	126.9	123.7	120.6	117.9	115.0	112.4

CYLINDER. Dia.	Stroke.	38	39	40	41	42	43	44	45	46	47	48	49	50	51	52	53	54	55	56
		DIAMETER OF DRIVERS.																		
15	24	142.1	138.4	135.0	131.8	128.6	125.6	122.8	120.0	117.3	115.0	112.4	110.0	108.0	106.0	103.9	101.9	100.0	98.4	96.5
16	20	....	131.0	128.0	124.9	121.9	119.0	116.2	113.9	111.1	108.9	106.7	104.4	102.4	100.3	98.5	96.6	94.9	93.1	91.4
16	22	....	....	....	....	134.0	130.9	127.9	125.0	122.4	119.8	117.3	114.8	112.6	110.2	108.1	106.0	104.1	102.1	100.5
16	24	....	....	....	....	146.0	142.8	139.5	137.3	133.5	130.2	128.0	125.2	122.9	120.1	118.0	115.9	113.7	111.6	109.7
17	20	....	....	....	....	137.6	134.4	131.1	128.4	125.6	122.9	120.4	117.9	115.6	113.3	111.1	109.1	107.0	105.0	103.2
17	22	....	....	....	....	151.2	147.9	144.7	141.2	138.2	135.3	132.5	129.9	127.1	124.5	122.1	120.0	117.8	115.6	113.6
17	24	....	....	....	....	165.1	161.1	157.5	154.0	150.6	147.3	144.5	141.3	138.7	135.8	133.0	130.7	128.0	126.0	123.8
18	22	....	....	....	....	169.7	165.7	162.0	158.2	154.8	151.5	148.3	145.4	142.5	139.9	137.0	134.4	132.0	129.7	127.1
18	24	....	....	....	....	185.1	181.0	175.5	172.7	169.0	165.2	161.9	158.7	155.5	152.4	149.5	146.7	144.0	141.2	138.9
18	26	....	....	....	....	200.5	196.0	191.5	187.1	183.1	179.2	175.5	172.0	168.4	165.0	162.0	158.9	156.0	153.1	150.4
19	22	....	....	....	....	189.0	184.6	180.3	176.4	172.5	168.9	165.3	162.0	158.8	155.6	152.8	149.8	147.0	144.3	141.8
19	24	....	....	....	....	206.2	201.5	196.9	192.5	188.3	184.2	180.5	176.8	173.2	170.0	166.5	163.4	160.3	157.5	154.8
19	26	....	....	....	....	223.2	218.0	213.0	208.6	203.9	199.4	195.4	191.3	187.7	184.0	180.4	177.4	173.9	170.6	167.6
20	24	....	....	....	....	228.5	223.5	218.1	213.3	208.5	204.2	200.0	196.0	192.0	188.2	184.6	181.0	177.8	174.5	171.3
20	26	....	....	....	....	248.0	242.0	236.5	231.2	226.2	221.5	216.6	212.1	208.0	204.0	200.0	196.5	192.7	189.3	185.9
21	22	....	....	....	....	231.0	226.0	220.5	216.0	210.5	206.3	202.0	198.0	194.0	190.2	186.5	183.0	179.6	176.5	173.2
21	24	....	....	....	....	252.0	246.1	239.8	235.2	230.9	225.1	220.5	216.0	211.6	207.5	203.5	199.6	196.0	192.5	189.0
21	26	....	....	....	....	272.8	266.6	260.5	254.8	249.2	243.0	238.8	234.0	229.3	225.0	220.5	216.5	212.5	208.5	205.0
22	24	....	....	....	....	276.5	270.0	263.9	258.0	252.3	247.0	242.0	236.8	232.0	227.5	223.4	219.0	215.0	211.2	207.3
22	26	....	....	....	....	299.5	292.6	286.0	279.5	273.5	267.4	262.0	256.2	251.6	246.4	241.7	237.3	232.6	229.5	224.6
22	28	....	....	....	....	322.6	315.1	308.0	301.1	294.6	288.3	282.3	276.5	271.0	265.7	260.6	255.6	250.9	246.4	242.0
22	30	....	....	....	....	345.7	337.6	330.0	322.6	315.6	308.9	302.5	296.3	290.4	284.7	279.2	273.9	268.8	264.0	259.2

CYLINDER. Dia.	Stroke.	76	77	78	79	80	81	82	83	84	85	86	87	88	89	90
		DIAMETER OF DRIVERS.														
15	24	71.0	70.1													
16	20	67.4	66.5	65.6												
16	22	74.1	73.2	72.3	71.4											
16	24	80.9	79.8	78.8	77.8	76.8										
17	20	76.0	75.0	74.1	73.1	72.2	71.3									
17	22	83.6	82.5	81.4	80.5	79.5	78.5	77.5								
17	24	91.2	90.0	88.9	87.7	86.7	85.6	84.5	83.5	82.5	81.5	80.6	79.6	78.8		
18	22	93.9	92.6	91.5	90.3	89.1	88.0	87.0	86.0	85.0	84.0	83.0	82.0	81.1	80.2	
18	24	102.1	100.9	99.7	98.5	97.2	96.0	94.9	93.7	92.6	91.5	90.4	89.4	88.4	87.5	
18	26	110.8	109.3	107.9	106.5	105.3	104.0	102.7	101.4	100.0	99.1	98.0	96.9	95.8	94.6	

TABLE 6.

CYLINDER.		DIAMETER OF DRIVERS.																		
Dia.	Stroke.	45	46	47	48	49	50	51	52	53	54	55	56	57	58	59	60	61	62	63
8	12																			
9	12	21.6																		
9	16	28.8	26.2	27.6	27.0															
10	18	40.0	39.1	38.3	37.2															
11	16	43.0	42.0	41.2	40.3	39.5	38.7	38.0												
11	18	48.4	47.3	46.3	45.3	44.4	43.5	42.6	41.8	41.0										
12	16	51.1	50.0	49.0	48.0	47.0	46.0	45.1	44.3	43.5	42.6	41.8								
12	18	57.6	56.3	55.0	54.0	52.8	51.8	50.8	49.8	48.8	48.0	47.1	46.3	45.4						
12	20	64.0	62.6	61.2	60.0	58.7	57.6	56.5	55.4	54.4	53.4	52.4	51.4	50.5	49.6					
12	22	70.5	68.9	67.4	66.0	64.6	63.3	62.1	61.0	59.8	58.7	57.6	56.5	55.5	54.6	53.7	52.8			
13	18	67.6	66.1	64.7	63.4	62.0	60.8	59.6	58.5	57.4	56.3	55.3	54.3	53.3	52.4	51.5	50.7	49.9		
13	20	75.1	73.5	72.0	70.5	69.0	67.6	66.3	65.0	63.8	62.6	61.4	60.3	59.3	58.3	57.3	56.3	55.4	54.5	53.
13	22	82.6	80.8	79.1	77.5	75.8	74.4	72.9	71.5	70.1	68.8	67.5	66.4	65.2	64.0	63.0	61.9	60.9	60.0	59.
14	18	78.5	76.7	75.1	73.5	72.0	70.5	69.2	67.9	66.5	65.4	64.2	63.0	61.9	60.8	59.8	58.8	57.8	56.9	56.
14	20	87.2	85.3	83.5	81.6	80.0	78.5	77.0	75.4	74.0	72.6	71.4	70.1	68.8	67.6	66.5	65.5	64.3	63.2	62.
14	22	95.9	93.9	91.8	90.0	88.0	86.2	84.6	83.0	81.4	80.0	78.5	77.0	75.6	74.4	73.1	71.9	70.7	69.6	68.
14	24	104.5	102.1	100.0	98.0	96.0	94.0	92.3	90.4	88.7	87.0	85.5	84.0	82.5	81.0	79.7	78.3	77.1	75.8	74.
15	18	90.1	88.1	86.3	84.5	82.7	81.0	79.5	78.0	76.5	75.0	73.7	72.4	71.0	69.9	68.7	67.5	66.4	65.3	64.
15	20	100.0	97.8	95.7	93.7	91.8	90.0	88.3	86.5	84.9	83.4	81.8	80.4	78.9	77.3	76.3	75.0	73.8	72.6	71.
15	22	110.0	107.5	105.3	103.0	101.1	99.0	97.1	95.2	93.4	91.7	90.0	88.5	86.8	85.4	84.0	82.5	81.2	79.8	78.

CYLINDER.		DIAMETER OF DRIVERS.																		
Dia.	Stroke	57	58	59	60	61	62	63	64	65	66	67	68	69	70	71	72	73	74	75
15	24	94.8	93.2	91.6	90.0	88.5	87.1	85.8	84.3	83.0	81.8	80.6	79.4	78.3	77.1	76.0	75.0	74.0	73.0	72.
16	20	89.8	88.4	86.9	85.4	84.0	82.6	81.3	80.0	78.8	77.6	76.5	75.3	74.2	73.1	72.1	71.1	70.1	69.2	68.
16	22	98.6	97.2	95.5	94.0	92.4	90.9	89.5	88.0	86.7	85.4	84.1	82.9	81.6	80.4	79.4	78.3	77.2	76.2	75.
16	24	107.8	105.9	104.6	102.4	100.7	99.0	97.6	96.0	94.6	93.2	91.8	90.4	89.2	87.9	86.6	85.4	84.2	83.1	82.
17	20	101.4	99.6	97.9	96.3	94.7	93.2	91.7	90.3	88.0	87.1	86.2	85.0	83.7	82.5	81.4	80.2	79.1	78.1	77.
17	22	111.6	109.6	107.8	105.9	104.1	102.6	101.0	99.5	98.0	96.5	95.0	93.5	92.2	90.8	89.5	88.3	87.1	86.9	84.
17	24	121.8	119.8	117.5	115.6	113.6	111.8	110.0	108.1	106.6	105.0	103.2	101.9	100.1	99.0	98.0	96.4	95.0	93.6	92.
18	22	125.0	122.9	120.9	118.8	116.9	115.0	113.1	111.3	109.8	108.1	106.4	104.9	103.2	101.8	100.1	99.0	97.6	96.4	95.
18	24	136.4	134.0	131.9	129.6	127.4	125.2	123.1	121.4	119.5	117.8	116.0	114.1	112.7	111.0	109.4	108.0	106.2	105.0	103.
18	26	147.8	145.1	142.8	140.4	138.0	135.9	133.7	131.7	129.7	127.6	125.7	123.9	122.0	120.3	118.7	117.0	115.3	113.8	112.
19	22	139.1	136.9	134.5	132.3	130.0	128.0	126.0	123.9	122.1	120.3	118.1	116.7	114.9	113.4	111.8	110.3	108.7	107.1	105.
19	24	152.0	149.3	147.0	144.4	142.0	139.8	137.5	135.3	133.1	131.2	129.2	127.4	125.6	123.8	122.0	120.1	118.7	117.0	115.
19	26	164.8	161.8	159.0	156.2	153.9	151.6	148.9	146.7	144.3	142.0	140.0	138.0	135.9	134.0	132.0	130.1	128.4	126.8	125.
20	24	168.9	166.5	162.8	160.0	157.4	154.8	152.1	150.0	147.5	145.4	143.1	141.0	139.0	137.1	135.1	133.1	131.3	129.7	127.
20	26	182.5	179.4	176.5	173.3	170.6	168.0	165.1	162.5	160.0	157.8	155.1	153.0	150.8	148.7	146.5	144.5	142.6	140.5	138.
21	22	170.1	167.2	164.8	161.7	159.1	156.4	154.0	151.5	149.1	147.0	144.8	142.8	140.6	138.6	136.6	134.8	132.8	131.0	129.
21	24	186.0	182.5	179.5	176.4	173.5	170.8	168.0	165.3	162.8	160.3	158.0	155.6	153.2	151.2	149.0	146.9	145.0	142.9	141.
21	26	201.1	197.7	194.6	191.3	188.1	185.0	182.1	179.1	176.5	173.8	171.2	168.8	165.2	163.8	161.6	159.7	157.1	155.0	152.
22	24	203.7	200.0	196.8	193.6	190.2	187.2	184.2	181.3	178.6	175.8	173.3	170.8	168.2	165.8	163.5	161.0	158.9	156.8	154.
22	26	220.6	217.0	215.2	209.7	206.6	203.0	199.6	196.4	193.5	190.4	187.5	185.0	182.3	179.7	177.0	174.5	172.2	169.0	167.
22	28	237.7	233.0	229.6	225.8	222.1	218.5	215.1	211.7	208.4	205.3	202.2	199.2	196.4	193.6	190.8	188.2	185.6	183.1	180.
22	30	254.7	250.3	246.1	242.0	238.0	234.2	230.4	226.8	223.3	220.0	216.7	213.5	210.4	207.4	204.5	201.6	198.9	196.2	193.

CYLINDER.		DIAMETER OF DRIVERS.														
Dia.	Stroke	76	77	78	79	80	8.	82	83	84	85	86	87	88	89	90
19	22	104.2	103.1	101.8	100.4	99.3	98.1	96.9	95.6	94.5	93.4	92.3	91.3	90.2	89.2	
19	24	113.0	112.4	111.0	109.6	108.3	106.9	105.5	104.3	103.0	101.9	100.7	99.6	98.5	97.4	
19	26	123.4	121.9	120.3	118.8	117.3	115.8	114.4	113.0	111.8	110.3	108.6	107.9	106.7	105.3	
20	24	126.1	124.4	123.0	121.5	120.0	118.4	117.0	115.5	114.1	112.8	111.5	110.1	109.0	107.8	
20	26	136.9	135.0	133.4	131.8	130.0	128.5	126.9	125.3	123.8	122.4	120.9	119.5	118.2	117.0	
21	22	127.5	126.0	124.2	122.9	121.2	119.8	117.4	116.9	115.3	114.0	112.8	111.4	110.0	109.0	
21	24	139.2	137.3	135.5	133.9	132.3	130.7	129.0	127.4	125.9	124.3	123.0	121.7	120.2	118.8	
21	26	150.9	149.0	147.0	145.1	143.3	141.7	140.0	138.1	136.6	134.9	133.4	131.9	130.4	128.8	
22	24	152.6	150.5	148.7	146.9	145.2	143.2	141.5	139.9	138.1	136.5	134.9	133.3	131.9	130.5	
22	26	165.3	163.2	161.0	159.0	157.3	155.3	153.2	151.4	149.7	147.8	146.1	144.5	142.9	141.2	

TABLE 7.

HEATING SURFACE OF FLUES IN SQUARE FEET.

OUTSIDE DIAMETER.

Outside Diameter.	Circumference.	LENGTH—FEET. 7	8	9	10	11	12	13	14	15	INCHES. 1	2	3	4	5	6	7	8	9	10	11	12	FRACTIONS. ¼	½	¾
1½ in.	4.71	2.74	3.13	3.53	3.92	4.31	4.70	5.11	5.49	5.88	.03	.06	.09	.13	.16	.19	.22	.26	.29	.32	.36	.39	.01	.01	.02
1¾ "	5.49	3.20	3.66	4.12	4.58	5.04	5.49	5.95	6.41	6.87	.03	.07	.11	.15	.19	.22	.26	.30	.34	.38	.41	.45	.01	.02	.03
2 "	6.28	3.66	4.18	4.71	5.23	5.76	6.28	6.80	7.33	7.85	.04	.08	.13	.17	.21	.26	.30	.34	.39	.43	.47	.52	.01	.02	.03
2¼ "	7.06	4.12	4.71	5.29	5.88	6.47	7.06	7.65	8.24	8.83	.04	.09	.14	.19	.24	.29	.34	.39	.44	.49	.53	.58	.01	.02	.03
2½ "	7.85	4.58	5.23	5.89	6.54	7.19	7.85	8.50	9.16	9.81	.05	.11	.16	.21	.27	.32	.38	.43	.49	.54	.59	.65	.01	.02	.04

Table No. 7 is given as a matter of convenience for obtaining the heating surface of tubes of the several diameters and lengths given therein.

For instance, to calculate the heating surface of say 253 2-inch diameter flues 12 feet 2¼ inches long, add together the figures in the vertical columns for the lengths corresponding to the outside diameter of the flue, this will give its heating surface and multiplied by the number of flues will give the total; as follows:

2 inches diameter, column 12 feet.......................6.28 sq. ft.
2 " " " 2 inches....................... .08 " "
2 " " " ¼ inch....................... .01 " "

6.37 " " x 253=1611.61 square feet for the 253 flues.

WEIGHT ON DRIVERS.

The proportion of the total weight of the locomotive that is carried on its driving wheels, varies with the different types.

In the ordinary eight-wheel American type the proportion of the total weight of the engine carried on its driving wheels, is about 64 per cent. when the fire-box extends down between the two axles, but when the back driving axle is located *under* the fire-box it is usually about 68per cent. of its total weight.

In the ten-wheel locomotive where the back driving axle is located *behind* the fire-box, the weight on the drivers is about 74 per cent., and when the back axle is *under* the fire-box it is about 78 per cent of the total weight of the engine.

In the Mogul engine where the back driving axle is located behind the fire-box the proportion carried on the driving wheels is about 83 per cent., but when the back axle is under the fire-box it will be about 86 per cent. of the total weight of the engine.

The consolidation type of engine has about 86 per cent. of its total weight on the drivers, when *two* of the driving axles are located under the fire-box, but when the back axle only is under the fire-box, it is about 89 per cent.

For convenience these figures are arranged in tabular form :

Kind of Locomotive.	Proportion of the total weight of the locomotive.	
	On the drivers.	On the truck.
Eight-wheel American type with fire-box between the two axles.	64 per cent.	36 per cent.
Eight-wheel with the back axle under the fire-box.	68 "	32 "
Ten-wheel with fire-box between back and middle axles.	74 "	26 "
Ten-wheel with fire-box over the back axle.	78 "	22 "
Mogul with fire-box between back and middle axle.	83 "	17 "
Mogul with fire-box over the back axle.	86 "	14 "
Consolidation with fire-box over the two rear axles.	86 "	14 "
Consolidation with fire-box over the one back axle only.	89 "	11 "

The table on the opposite page which gives the resistance in pounds per ton (of 2000 lbs.) is taken from Forney's Catechism of the Locomotive and was calculated by the rules given on a former page. The various speeds are indicated in the heading at the top of the columns, and the rate of gradients (rise in feet per mile) is given in the first column on the left.

TABLE OF RESISTANCES OF RAILROAD TRAINS,

ON A STRAIGHT TRACK,

WITH DIFFERENT GRADES AND SPEEDS.

Rise of gradient, feet per mile.	Resistance due to ascent alone in pounds per ton (2000 lbs.) of train.	Total resistance, pounds per ton, at rate of 5 miles per hour.	10 miles per hour.	15 miles per hour.	20 miles per hour.	25 miles per hour.	30 miles per hour.	35 miles per hour.	40 miles per hour.	45 miles per hour.	50 miles per hour.	60 miles per hour.	70 miles per hour.
0		6.1	6.6	7.3	8.3	9.6	11.2	13.1	15.3	17.8	20.6	27.0	31.6
5	1.8	7.9	8.4	9.1	10.1	11.4	13.0	14.9	17.1	19.6	22.4	28.8	36.4
10	3.7	9.8	10.3	11.0	12.0	13.4	14.9	16.8	19.0	21.5	24.3	30.7	38.3
15	5.6	11.7	12.2	12.9	13.9	15.2	16.8	18.7	21.9	24.4	27.2	33.6	41.2
20	7.5	13.6	14.1	14.8	15.8	17.1	18.7	20.6	22.8	25.3	28.1	34.5	42.1
25	9.4	15.5	16.0	16.7	17.7	19.0	20.6	22.5	24.7	27.2	31.0	37.4	45.0
30	11.3	17.4	17.9	18.6	19.6	21.9	22.5	24.4	26.6	29.1	31.9	38.3	45.9
35	13.2	19.3	19.8	20.5	21.5	22.8	24.4	26.3	28.5	31.0	33.8	40.2	47.8
40	15.1	21.2	21.7	22.4	23.4	24.7	26.3	28.2	30.4	32.9	35.7	42.1	49.7
45	17.0	23.1	23.6	24.3	25.3	26.6	28.2	30.1	32.3	34.8	37.6	44.0	51.6
50	18.9	25.0	25.5	26.2	27.2	28.5	30.1	32.0	34.2	36.7	39.5	45.9	53.5
60	22.7	28.8	29.3	30.0	31.0	32.3	33.9	35.8	38.0	40.5	43.3	49.9	57.3
70	26.5	32.6	33.1	33.8	34.8	36.1	37.7	39.6	41.8	44.3	47.1	53.5	61.1
80	30.3	36.4	36.9	37.6	38.6	39.9	40.5	42.4	44.6	47.1	49.9	56.3	63.9
90	34.0	41.0	40.6	41.3	42.3	43.6	45.2	47.1	49.3	51.8	54.6	61.0	68.6
100	37.8	43.9	44.4	45.1	46.1	47.4	49.0	51.9	54.1	56.6	59.4	65.8	73.1
110	41.6	47.7	48.2	48.9	49.9	51.2	52.8	54.7	56.9	59.1	62.2	68.6	76.2
120	45.4	51.5	52.0	52.7	53.7	55.0	56.6	58.5	60.7	63.2	66.0	72.4	80.0
130	49.2	55.3	55.8	56.5	57.5	58.8	60.4	62.3	64.5	67.0	69.8	76.2	83.8
140	53.0	59.1	59.6	60.3	61.3	62.6	64.2	66.1	68.3	70.8	73.6	80.0	87.6
150	56.8	62.9	63.4	64.1	65.1	66.4	68.0	69.9	72.1	74.6	77.4	83.8	91.4
160	60.6	66.7	67.2	67.9	68.9	70.2	71.8	73.7	75.9	78.4	81.2	87.6	95.2
170	64.3	70.4	70.9	71.6	72.6	73.9	75.5	77.4	79.6	82.1	84.9	91.3	98.9
180	68.1	74.2	74.7	75.4	76.4	77.7	79.3	81.2	83.4	85.9	88.7	95.1	102.7
190	71.9	78.0	78.5	79.2	80.2	81.5	83.1	85.0	87.2	89.7	92.5	98.9	106.5
200	75.7	81.8	82.3	83.0	84.0	85.3	86.9	88.8	91.0	93.5	96.3	102.7	110.3
210	79.5	85.6	86.1	86.8	87.8	89.1	90.7	92.6	94.8	97.3	100.1	106.5	114.1
220	83.3	89.4	89.9	90.6	91.6	92.9	94.5	96.4	98.6	101.1	103.9	110.3	117.6
230	87.1	93.2	93.7	94.4	95.4	96.7	98.3	100.2	102.4	104.9	107.7	114.1	121.7
240	90.8	96.9	97.4	98.1	99.1	100.4	102.0	103.9	106.1	108.6	111.4	117.8	125.4
250	94.6	100.7	101.2	101.9	102.9	103.2	105.8	107.7	109.9	112.4	113.2	121.6	129.2
260	98.4	104.5	105.0	105.7	106.7	107.0	108.6	110.5	112.7	115.2	118.0	124.4	132.0
270	102.2	108.3	108.8	109.5	110.5	111.8	113.4	115.3	117.5	120.0	122.8	129.2	136.8
280	106.0	112.1	112.6	113.3	114.3	115.6	117.2	119.1	121.3	123.8	126.6	133.0	140.6
290	109.8	115.9	116.4	117.1	118.1	119.4	121.0	122.9	125.1	127.6	130.4	136.8	144.4
300	113.6	119.7	120.2	120.9	121.9	123.2	124.8	126.7	128.9	131.4	134.2	140.6	148.2

PLATES

—AND—

Tables of Dimensions and Capacity

—OF—

LOCOMOTIVES.

PLATES AND TABLES OF DIMENSIONS AND CAPACITY OF LOCOMOTIVES OF 4 FT., 8½ IN. GAUGE OR WIDER.

IN the following tables the principal dimensions, weight, etc., and the calculated capacity for hauling loads is given for the different classes of locomotives manufactured by the Rogers Locomotive Company.

The hauling capacity is based on the average piston pressure for the whole stroke as it is affected by changes in the speed of the piston, as explained in the preceding pages. At lower train speeds than those given in the Tables of Dimensions and Capacity, the piston speed will, of course, be less and consequently the pressure will be greater and the locomotive will haul a proportionally greater load than that given in the table

The calculations are made for straight lines, and for the grades and speeds specified in each table. An allowance which has been explained on a former page must be made for curves.

PLATE I.

Eight Wheel Standard Locomotives

FOR PASSENGERS.

Gauge, 4 ft., 8½ in., or wider. Fuel, Bituminous or Anthracite Coal.

General Design shown by Plate I.

	Cylinders. Diameter and stroke. INCHES.	Dia'eter of Driving Wheels. INCHES.	Wheel Base. Of Driving Wheels.	Wheel Base. Total.	Weight, in running order. POUNDS. On Driving Wheels. LBS.	On Truck. LBS.	Total. LBS.	Separate Tender. Capacity of Tank. GALS.
1	15 x 24	56	8 ft. 3 in.	22 ft. 3 in.	46,000	26,000	72,000	2,000
2	16 x 24	62	8 ft. 3 in.	22 ft. 4 in.	52,000	28,000	80,000	2,500
3	17 x 24	62	8 ft. 6 in.	22 ft. 9 in.	57,000	32,000	89,000	3,000
4	18 x 24	62	8 ft. 9 in.	23 ft. 2 in.	62,000	34,000	96,000	3,000
5	18 x 24	66	8 ft. 0 in.	22 ft. 9 in.	77,000	35,000	112,000	3,500

Load in tons of 2,000 pounds in addition to Engine and Tender, at 30 to 40 miles an hour, on a grade per mile of

	On a Level.	10 ft.	20 ft.	40 ft.	60 ft.	80 ft.	100 ft.	125 ft.	150 ft.
1	864	628	486	330	241	191	147	113	87
2	864	628	486	330	241	191	147	113	87
3	1,000	724	560	383	279	223	170	128	100
4	1,061	777	602	404	294	233	178	135	105
5	1,076	780	605	406	296	235	178	135	125

PLATE II.

Eight Wheel Standard Locomotives

FOR PASSENGERS.

Gauge, 4 ft., 8½ in., or wider. Fuel, Bituminous or Anthracite Coal.

General Design shown by Plate II.

	Cylinders. Diameter and Stroke.	Dia'eter of Driving Wheels.	Wheel Base.		Weight, in running order POUNDS.			Separate Tender.
			Of Driving Wheels.	Total.	On Driving Wheels.	On Truck.	Total.	Capacity of Tank.
	INCHES.	INCHES.			LBS.	LBS.	LBS.	GALS.
1	15 x 24	66	8 ft. 3 in.	22 ft. 3 in.	48,000	27,000	75,000	2,000
2	16 x 24	66	8 ft. 3 in.	22 ft. 3 in.	53,000	30,000	83,000	2,500
3	17 x 24	66	8 ft 6 in.	22 ft. 9 in	59,000	33,000	92,000	3,000
4	18 x 24	66	9 ft. 0 in.	23 ft. 4 in.	64,000	36,000	100,000	3,300
5	18 x 26	69	9 ft. 1 in.	23 ft 6 in.	69,000	39,000	108,000	4,000

Load in tons of 2 000 pounds in addition to Engine and Tender, at 30 to 50 miles an hour, on a grade per mile of

	On a Level.	10 ft.	20 ft.	40 ft	60 ft.	80 ft	100 ft	125 ft.	150 ft.
1	466	360	276	176	122	94	64	44	28
2	840	595	462	306	220	172	129	97	70
3	884	645	499	332	241	189	143	107	81
4	972	728	562	375	272	215	162	122	93
5	1,086	788	614	410	297	237	178	135	103

PLATE III.

Eight Wheel Standard Locomotives

FOR PASSENGERS.

Gauge, 4 ft. 8½ in., or wider. Fuel, Bituminous Coal.

General Design shown by Plate III.

	Cylinders. Diameter and Stroke.	Diam'ter of Driving Wheels	Wheel Base.		Weight, in running order. POUNDS.			Separate Tender.
	INCHES.	INCHES.	Of Driving Wheels.	Total.	On Driving Wheels. LBS.	On Truck. LBS.	Total. LBS.	Capacity of Tank. GALS.
1	17 x 24	66	8 ft. 9 in.	23 ft. 0 in.	62,000	34,000	96,000	3,000
2	18 x 24	66	9 ft. 0 in.	23 ft. 6 in.	67,000	38,000	105,000	3,500
3	19 x 24	66	9 ft. 1 in.	23 ft. 9 in.	72,000	40,000	112,000	3,500
4	19 x 26	69	8 ft. 0 in.	23 ft. 0 in.	82,000	38,000	120,000	4,000
5								

Load in tons of 2,000 pounds in addition to Engine and Tender, at 45 miles an hour, on a grade per mile of

	On a Level.	10 ft.	20 ft.	40 ft.	60 ft.	80 ft.	100 ft.	125 ft.	150 ft.
1	555	445	366	262	231	159	118	92	70
2	628	505	416	300	230	184	138	116	94
3	706	569	470	339	262	211	160	128	100
4	730	588	486	350	272	218	165	132	102
5									

PLATE IV.

Eight Wheel Standard Locomotives

FOR PASSENGERS.

Gauge, 4 ft. 8½ in., or wider. Fuel, Bituminous or Anthracite Coal.

General Design shown by Plate IV.

	Cylinders. Diameter and Stroke.	Diam'ter of Driving Wheels.	Wheel Base.		Weight, in running order. POUNDS.			Separate Tender.
			Of Driving Wheels.	Total.	On Driving Wheels.	On Truck.	Total.	Capacity of Tank.
	INCHES.	INCHES			LBS.	LBS.	LBS.	GALS.
1	17 x 24	62	8 ft. 0 in.	21 ft. 10 in.	59,000	28,000	87,000	2,500
2	18 x 24	62	8 ft. 6 in.	22 ft. 10 in.	65,000	31,000	96,000	3,000
3	19 x 24	68	8 ft. 6 in.	22 ft. 10 in.	75,000	35,000	110,000	3,500
4	19 x 24	68	7 ft. 6 in.	21 ft. 8 in.	80,000	29,000	109,000	3,500
5	19 x 26	78	8 ft. 6 in.	23 ft. 8 in.	86,000	42,000	128,000	4,000

	Load in tons of 2,000 pounds in addition to Engine and Tender, at 45 miles an hour, on a grade per mile of								
	On a Level.	10 ft.	20 ft.	40 ft.	60 ft.	80 ft.	100 ft.	125 ft.	150 ft.
1	602	485	400	290	225	180	137	110	86
2	670	548	454	330	258	208	158	128	100
3	670	548	454	330	258	208	158	128	100
4	670	548	454	330	258	208	158	128	100
5	670	548	454	330	258	208	158	128	100

PLATE V.

Eight Wheel Standard Locomotives

FOR PASSENGERS.

Gauge, 4 ft. 8½ in., or wider. Fuel, Bituminous Coal.

General Design shown by Plate V.

	Cylinders. Diameter and Stroke. INCHES.	Diameter of Driving Wheels. INCHES.	Wheel Base. Of Driving Wheels.	Wheel Base. Total.	Weight, in running order. POUNDS. On Driving Wheels. LBS.	On Truck. LBS.	Total. LBS.	Separate Tender. Capacity of Tank. GALS.
1	18 x 24	68	9 ft. 1 in.	23 ft. 5 in.	70,000	40,000	110,000	3,800
2	19 x 24	68	7 ft. 6 in.	21 ft. 10 in.	76,000	36,000	112,000	4,000
3	19 x 26	78	8 ft. 6 in.	22 ft. 6 in.	80,000	36,000	116,000	4,000
4								
5								

Load in tons of 2,000 pounds in addition to Engine and Tender, at 50 miles an hour, on a grade per mile of

	On a Level.	10 ft.	20 ft.	40 ft.	60 ft.	80 ft.	100 ft.	125 ft.	150 ft.
1	524	419	352	257	195	159	119	93	72
2	590	476	400	295	226	184	140	112	88
3	590	476	400	295	226	184	140	112	88
4									
5									

PLATE VI.

Eight Wheel Standard Locomotives

FOR PASSENGERS.

Gauge, 4 ft. 8½ in., or wider. Fuel, Bituminous Coal.

General Design shown by Plate VI.

	Cylinders. Diameter and Stroke.	Diam'ter of Driving Wheels.	Wheel Base.		Weight, in running order. POUNDS.			Separate Tender.
			Of Driving Wheels.	Total.	On Driving Wheels.	On Truck.	Total.	Capacity of Tank.
	INCHES.	INCHES.			LBS.	LBS.	LBS.	GALS.
1	18 x 24	69	8 ft. 6 in.	24 ft. 0 in.	65,500	36,500	102,000	3,600
2	19 x 24	69	8 ft. 6 in.	24 ft. 0 in.	70,000	38,000	108,000	3,000
3	19 x 26	72	8 ft. 6 in.	24 ft. 0 in.	78,000	39,000	117,000	4,200
4								
5								

Load in tons of 2,000 pounds in addition to Engine and Tender, at 40 to 50 miles an hour, on a grade per mile of

	On a Level.	10 ft.	20 ft.	40 ft.	60 ft.	80 ft.	100 ft.	125 ft.	150 ft.
1	972	728	562	375	272	215	162	122	93
2	1,109	814	628	420	305	241	183	139	107
3	1,157	840	654	438	317	250	191	144	110
4									
5									

PLATE VII.

Eight Wheel Standard Locomotives

FOR PASSENGERS.

Gauge, 4 ft., 8½ in., or wider. Fuel, Bituminous Coal.

General Design shown by Plate VII.

	Cylinders. Diameter and Stroke. INCHES.	Dia'eter of Driving Wheels INCHES.	Wheel Base. Of Driving Wheels.	Wheel Base. Total.	Weight, in running order. POUNDS. On Driving Wheels. LBS.	Weight, in running order. On Truck. LBS.	Weight, in running order. Total. LBS.	Separate Tender. Capacity of Tank. GALS.
1	18 x 24	68	9 ft. 0 in.	23 ft. 4 in.	67,000	37,000	104,000	3,500
2	18 x 24	72	9 ft. 0 in.	23 ft. 4 in.	69,000	38,000	107,000	3,500
3	19 x 24	72	9 ft. 1 in.	23 ft. 6 in.	73,000	39,000	112,000	4,000
4	19 x 26	72	8 ft. 0 in.	22 ft. 6 in.	80,000	38,000	118,000	4,000
5								

Load in tons of 2,000 pounds in addition to Engine and Tender, at 40 to 50 miles an hour, on a grade per mile of

	On a Level.	10 ft.	20 ft.	40 ft.	60 ft.	80 ft.	100 ft.	125 ft.	130 ft.
1	972	728	562	375	272	215	162	122	93
2	907	612	410	339	245	191	127	107	81
3	1,000	726	568	378	273	214	143	120	91
4	1,157	840	654	438	317	250	191	144	110
5									

PLATE VIII.

Eight Wheel Standard Locomotives

FOR FREIGHT.

COMPOUND FOR PASSENGERS.

Gauge, 4 ft. 8½ in., or wider. Fuel, Bituminous or Anthracite Coal.

General Design shown by Plate VIII.

	Cylinders. Diameter and Stroke. INCHES.	Diam'ter of Driving Wheels. INCHES.	Wheel Base. Of Driving Wheels.	Wheel Base. Total.	Weight, in running order. POUNDS. On Driving Wheels. LBS.	On Truck. LBS.	Total. LBS.	Separate Tender. Capacity of Tank. GALS.
1	15 x 22	54	6 ft. 0 in.	19 ft. 0 in.	48,000	22,000	70,000	2,000
2	16 x 24	54	6 ft. 6 in.	20 ft. 3 in.	53,000	24,000	77,000	2,500
3	17 x 24	54	7 ft. 0 in.	20 ft. 10 in.	58,000	27,000	85,000	3,000
4	18 x 24	54	7 ft. 6 in.	21 ft. 10 in.	65,000	31,000	96,000	3,500
*5	Compound, H.P. 14x24, L.P. 24x24	66	8 ft. 3 in.	22 ft. 11 in.	72,000	44,000	116,000	3,600

Load in tons of 2,000 pounds in addition to Engine and Tender, at 15 miles an hour, on a grade per mile of

	On a Level.	10 ft.	20 ft.	40 ft.	60 ft.	80 ft.	100 ft.	125 ft.	150 ft.
1	1,770	1,155	843	537	385	296	236	188	148
2	2,040	1,332	972	619	445	344	273	214	172
3	2,292	1,496	1,093	696	501	385	308	241	194
4	2,581	1,684	1,230	785	565	434	348	273	220
*5	1,167	847	607	442	319	253	192	146	112

* Compound Passenger Locomotive at 40 miles per hour.

PLATE 2A.

Eight Wheel Standard Locomotives

FOR PASSENGERS OR FREIGHT.

Gauge, 4 ft., 8½ in., or wider. Fuel, Bituminous Coal.

General Design shown by Plate IX.

	Cylinders. Diameter and Stroke. INCHES.	Dia'eter of Driving Wheels. INCHES.	Wheel Base. Of Driving Wheels.	Wheel Base. Total.	Weight, in running order. POUNDS. On Driving Wheels. LBS.	On Truck. LBS.	Total. LBS.	Separate Tender. Capacity of Tank. GALS.
1	17 x 24	62	8 ft. 6 in.	22 ft. 9 in.	58,000	31,000	89,000	3,000
2	18 x 24	62	8 ft. 9 in.	23 ft. 0 in.	63,000	34,000	97,000	3,500
*3	17 x 24	72	8 ft. 6 in.	22 ft. 11 in.	60,000	36,000	96,000	3,500
4								
5								

Load in tons of 2,000 pounds in addition to Engine and Tender, at 25 miles an hour, on a grade per mile of

	On a Level.	10 ft.	20 ft.	40 ft.	60 ft.	80 ft.	100 ft.	125 ft.	150 ft.
1	949	688	537	360	262	207	158	119	90
2	1,043	756	590	396	288	227	173	130	99
*3	940	682	532	336	259	203	154	117	90
4									
5									

* Passengers only.—45 miles per hour.

PLATE X.

Mogul Locomotives

FOR FREIGHT.

Gauge, 4 ft., 8½ in., or wider. Fuel, Bituminous Coal or Wood

General Design shown by Plate X.

	Cylinders. Diameter and Stroke. INCHES.	Dia'eter of Driving Wheels. INCHES.	Wheel Base. Of Driving Wheels.	Wheel Base. Total.	Weight, in running order. POUNDS. On Driving Wheels. LBS.	Weight, in running order. POUNDS. On Truck. LBS.	Weight, in running order. POUNDS. Total. LBS.	Separate Tender. Capacity of Tank. GALS.
1	17 x 22	57	14 ft. 6 in.	22 ft. 0 in.	71,000	16,000	87,000	2,500
2	17 x 24	57	15 ft. 6 in.	23 ft. 0 in.	75,000	17,000	92,000	3,000
3								
4								
5								

Load in tons of 2,000 pounds in addition to Engine and Tender, at 20 miles an hour, on a grade per mile of

	On a Level.	10 ft.	20 ft.	40 ft.	60 ft.	80 ft.	100 ft.	125 ft.	150 ft.
1	1,434	1,163	880	570	414	318	257	200	160
2	1,586	1,295	966	637	455	353	273	219	175
3									
4									
5									

PLATE XI.

Mogul Locomotives

FOR FREIGHT.

Gauge, 4 ft., $8\frac{1}{2}$ in., or wider. Fuel, Bituminous Coal.

General Design shown by Plate XI.

	Cylinders. Diameter and Stroke. INCHES.	Dia'eter of Driving Wheels. INCHES.	Wheel Base. Of Driving Wheels.	Wheel Base. Total.	Weight, in running order. POUNDS. On Driving Wheels. LBS.	Weight, in running order. POUNDS. On Truck. LBS.	Weight, in running order. POUNDS. Total. LBS.	Separate Tender. Capacity of Tank. GALS.
1	19 x 24	54	13 ft. 0 in.	20 ft. 8 in.	92,000	16,000	108,000	3,500
2								
3								
4								
5								

Load in tons of 2,000 pounds in addition to Engine and Tender, at 20 miles an hour, on a grade per mile of

	On a Level.	10 ft.	20 ft.	40 ft.	60 ft.	80 ft.	100 ft.	125 ft.	150 ft.
1	2,314	1,871	1,362	872	628	482	387	301	245
2									
3									
4									
5									

PLATE XII.

Mogul Locomotives

FOR FREIGHT.

Gauge, 4 ft. 8½ in., or wider   Fuel, Bituminous or Anthracite Coal.

General Design shown by Plate XII.

	Cylinders. Diameter and Stroke.	Diam'ter of Driving Wheels.	Wheel Base.		Weight, in running order. POUNDS.			Separate Tender.
			Of Driving Wheels.	Total.	On Driving Wheels.	On Truck.	Total.	Capacity of Tank.
	INCHES.	INCHES.			LBS.	LBS.	LBS.	GALS.
1	18 x 24	50	13 ft. 0 in.	20 ft. 6 in.	88,000	14,000	102,000	3,000
2	19 x 24	50	13 ft. 6 in.	22 ft. 2 in.	93,000	17,000	110,000	3,500
3	19 x 26	54	14 ft. 6 in.	22 ft. 2 in.	110,000	17,000	127,000	4,000
4								
5								

Load in tons of 2,000 pounds in addition to Engine and Tender, at 20 miles an hour, on a grade per mile of

	On a Level.	10 ft.	20 ft.	40 ft.	60 ft.	80 ft.	100 ft.	125 ft.	150 ft.
1	2,223	1,825	1,335	851	617	475	382	301	244
2	2,514	2,033	1,488	952	689	531	428	338	274
3	3,098	2,464	1,476	940	676	519	414	324	262
4									
5									

PLATE XIII

Mogul Locomotives

FOR FREIGHT.

Gauge, 4 ft., 8½ in., or wider. Fuel, Bituminous or Anthracite Coal.

General Design shown by Plate XIII.

	Cylinders. Diameter and Stroke. INCHES.	Dia'eter of Driving Wheels. INCHES.	Wheel Base. Of Driving Wheels.	Wheel Base. Total.	Weight, in running order. POUNDS. On Driving Wheels. LBS.	On Truck. LBS.	Total. LBS.	Separate Tender. Capacity of Tank. GALS.
1	19 x 26	63	14 ft. 0 in.	21 ft. 9 in.	109,000	17,000	126,000	3,500
2	19 x 26	56	14 ft. 0 in.	21 ft. 9 in.	108,000	17,000	125,000	4,000
3								
4								
5								

Load in tons of 2,000 pounds in addition to Engine and Tender, at 25 miles an hour, on a grade per mile of

	On a Level.	10 ft.	20 ft.	40 ft.	60 ft.	80 ft.	100 ft.	125 ft.	150 ft.
1	1,883	1,554	1,157	750	512	430	342	265	212
2	2,166	1,790	1,336	870	633	490	405	317	256
3									
4									
5									

PLATE XIV

Mogul Locomotives

FOR FREIGHT.

Gauge, 4 ft. $8\frac{1}{2}$ in., or wider. Fuel, Bituminous or Anthracite Coal.

General Design shown by Plate XIV.

	Cylinders. Diameter and Stroke. INCHES.	Diam'ter of Driving Wheels. INCHES.	Wheel Base. Of Driving Wheels.	Wheel Base. Total.	Weight, in running order. POUNDS. On Driving Wheels. LBS.	On Truck. LBS.	Total. LBS.	Separate Tender. Capacity of Tank. GALS.
1	18 x 24	54	13 ft. 10 in.	21 ft. 8 in.	88,000	17,000	105,000	3,000
2	18 x 24	62	13 ft. 10 in.	21 ft. 8 in.	89,000	17,000	106,000	3,000
3	19 x 24	54	13 ft. 10 in.	21 ft. 8 in.	99,000	17,000	116,000	3,500
4								
5								

Load in tons of 2,000 pounds in addition to Engine and Tender, at 20 miles an hour, on a grade per mile of

	On a Level.	10 ft.	20 ft.	40 ft.	60 ft.	80 ft.	100 ft.	125 ft.	150 ft.
1	2,194	1,536	1,150	746	563	432	346	271	218
2	1,916	1,555	1,134	720	516	395	315	243	197
3	2,314	1,870	1,361	871	627	481	386	303	244
4									
5									

PLATE XV.

Mogul Locomotives

FOR PASSENGERS OR FREIGHT.

Gauge, 4 ft., 8½ in., or wider. Fuel, Bituminous or Anthracite Coal.

General Design shown by Plate XV.

	Cylinders. Diameter and Stroke. INCHES.	Dia'eter of Driving Wheels. INCHES.	Wheel Base. Of Driving Wheels.	Wheel Base. Total.	Weight, in running order. POUNDS. On Driving Wheels. LBS.	On Truck. LBS.	Total. LBS.	Separate Tender. Capacity of Tank. GALS.
1	19 x 24	62	15 ft. 3 in.	23 ft. 4 in.	93,000	18,000	111,000	3,500
2								
3								
4								
5								

Load in tons of 2,000 pounds in addition to Engine and Tender, at 30 miles an hour, on a grade per mile of

	On a Level.	10 ft.	20 ft.	40 ft.	60 ft.	80 ft.	100 ft.	125 ft.	150 ft.
1	1,778	1,457	1,104	716	518	398	319	253	200
2									
3									
4									
5									

PLATE XVI.

Mogul Locomotives

FOR FREIGHT.

Gauge, 4 ft., 8½ in., or wider. Fuel, Bituminous Coal.

General Design shown by Plate XVI.

	Cylinders. Diameter and Stroke.	Dia'eter of Driving Wheels	Wheel Base.		Weight, in running order. POUNDS.			Separate Tender.
			Of Driving Wheels.	Total.	On Driving Wheels.	On Truck.	Total.	Capacity of Tank.
	INCHES.	INCHES.			LBS.	LBS.	LBS.	GALS.
1	19 x 26	56	14 ft. 0 in.	21 ft. 8 in.	107,000	19,000	126,000	3,800
2	20 x 26	54	14 ft 0 in.	21 ft. 8 in.	113,000	18,000	131,000	4,000
3								
4								
5								

	Load in tons of 2,000 pounds in addition to Engine and Tender, at 25 miles an hour, on a grade per mile of								
	On a Level.	10 ft.	20 ft.	40 ft.	60 ft.	80 ft.	100 ft.	125 ft.	150 ft.
1	1,821	1,478	1,148	757	556	432	347	272	220
2	2,030	1,713	1,322	884	654	510	411	328	267
3									
4									
5									

PLATE XVII.

Mogul Locomotives

COMPOUND

FOR FREIGHT.

Gauge, 4 ft., 8½ in., or wider. Fuel, Bituminous or Anthracite Coal.

General Design shown by Plate XVII.

	Cylinders. Diameter and Stroke. INCHES.	Dia'eter of Driving Wheels. INCHES.	Wheel Base. Of Driving Wheels.	Wheel Base. Total.	Weight, in running order. POUNDS. On Driving Wheels. LBS.	On Truck. LBS.	Total. LBS.	Separate Tender. Capacity of Tank. GALS.
1	H. P. 20 x 26. L. P. 29 x 26	56	14 ft. 0 in.	21 ft. 8 in.	107,500	20,500	128,000	3,800
2								
3								
4								
5								

Load in tons of 2,000 pounds in addition to Engine and Tender, at 25 miles an hour, on a grade per mile of

	On a Level.	10 ft.	20 ft.	40 ft.	60 ft.	80 ft.	100 ft.	125 ft.	150 ft.
1	1,821	1,478	1,148	737	556	432	347	272	220
2									
3									
4									
5									

PLATE XVIII.

Ten Wheel Locomotives

FOR FREIGHT.

Gauge, 4 ft. 8½ in., or wider Fuel, Bituminous Coal.

General Design shown by Plate XVIII.

	Cylinders. Diameter and Stroke.	Diam'ter of Driving Wheels.	Wheel Base.		Weight, in running order. POUNDS.			Separate Tender.
			On Driving Wheels.	Total.	On Driving Wheels.	On Truck.	Total	Capacity of Tank.
	INCHES.	INCHES.			LBS.	LBS.	LBS	GALS.
1	17 x 24	54	13 ft. 6 in.	23 ft. 11 in.	74,000	26,000	100,000	3,000
2	18 x 24	54	14 ft. 0 in.	24 ft. 5 in.	80,000	27,000	107,000	3,500
3	18 x 26	56	12 ft. 0 in.	22 ft. 3 in.	98,000	27,000	125,000	4,000
4								
5								

Load in tons of 2,000 pounds in addition to Engine and Tender, at 20 miles an hour, on a grade per mile of

	On a Level.	10 ft.	20 ft.	40 ft.	60 ft.	80 ft.	100 ft.	125 ft.	150 ft.
1	1,644	1,356	1,010	651	473	388	301	234	188
2	2,191	1,356	1,150	746	543	420	338	276	214
3	2,365	1,658	1,242	801	593	450	357	276	214
4									
5									

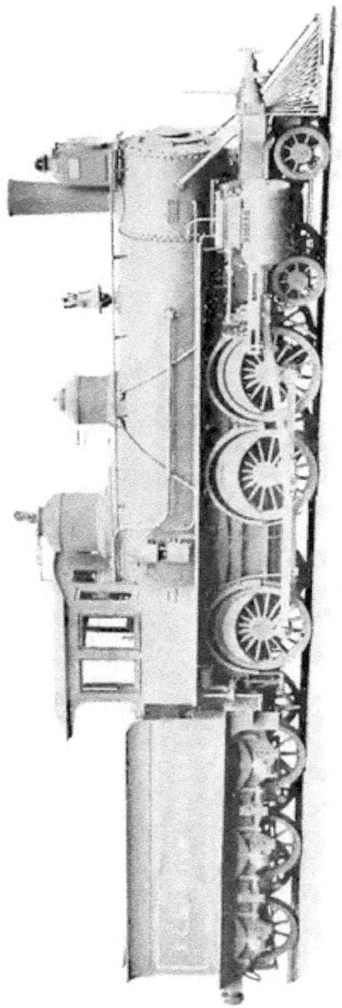

PLATE XIX.

Ten Wheel Locomotives

FOR FREIGHT.

Gauge, 4 ft., 8½ in., or wider. Fuel, Bituminous Coal.

General Design shown by Plate XIX.

	Cylinders, Diameter and Stroke.	Dia'eter of Driving Wheels.	Wheel Base.		Weight, in running order. POUNDS.			Separate Tender.
			Of Driving Wheels.	Total.	On Driving Wheels.	On Truck.	Total.	Capacity of Tank.
	INCHES.	INCHES.			LBS.	LBS.	LBS.	GALS.
1	18 x 24	56	13 ft. 4 in	23 ft. 8 in.	80,000	32,000	112,000	3,200
2	19 x 24	56	14 ft. 0 in.	24 ft. 4 in.	88,000	34,000	122,000	3,600
3								
4								
5								

	Load in tons of 2,000 pounds in addition to Engine and Tender, at 20 miles an hour, on a grade per mile of								
	On a Level.	10 ft.	20 ft.	40 ft.	60 ft.	80 ft.	100 ft.	125 ft.	150 ft.
1	2,481	1,959	1,178	749	536	410	327	255	204
2	2,769	2,200	1,316	837	601	460	367	286	229
3									
4									
5									

PLATE XX.

Ten Wheel Compound Locomotive

FOR FREIGHT.

Gauge, 4 ft. 8½ in., or wider. Fuel, Bituminous Coal.

General Design shown by Plate XX.

	Cylinders. Diameter and Stroke.	Diam'ter of Driving Wheels.	Wheel Base.		Weight, in running order. POUNDS.			Separate Tender.
			Of Driving Wheels.	Total.	On Driving Wheels.	On Truck.	Total	Capacity of Tank.
	INCHES.	INCHES.			LBS.	LBS.	LBS.	GALS.
1	H.P. 20 / L.P. 29 x 26	50	12 ft. 3 in.	22 ft. 9 in.	97,000	30,500	127,500	3,500
2								
3								
4								
5								

Load in tons of 2,000 pounds in addition to Engine and Tender, at 20 miles an hour, on a grade per mile of

	On a Level.	10 ft.	20 ft.	40 ft.	60 ft.	80 ft.	100 ft.	125 ft.	150 ft.
1	3,036	2,070	1,549	1,015	744	576	468	376	304
2									
3									
4									
5									

PLATE XXI.

Ten Wheel Locomotives

FOR FREIGHT.

Gauge, 4 ft., 8½ in., or wider. Fuel, Anthracite Coal.

General Design shown by Plate XXI.

	Cylinders. Diameter and Stroke. INCHES.	Dia'eter of Driving Wheels INCHES.	Wheel Base. Of Driving Wheels.	Wheel Base. Total.	Weight, in running order, POUNDS. On Driving Wheels. LBS.	On Truck. LBS.	Total. LBS.	Separate Tender. Capacity of Tank. GALS.
1	21 x 26	63	12 ft. 0 in.	22 ft. 11 in.	115,000	31,000	146,000	4,000
2								
3								
4								
5								

Load in tons of 2,000 pounds in addition to Engine and Tender, at 20 miles an hour, on a grade per mile of

	On a Level.	10 ft.	20 ft.	40 ft.	60 ft.	80 ft.	100 ft.	125 ft.	150 ft.
1	2,260	1,533	1,136	733	527	402	320	257	195
2									
3									
4									
5									

PLATE XXVII.

Ten Wheel Locomotives

FOR PASSENGERS OR FREIGHT.

Gauge, 4 ft. 8½ in., or wider. Fuel, Bituminous Coal.

General Design shown by Plate XXII.

	Cylinders. Diameter and Stroke.	Diam'ter of Driving Wheels.	Wheel Base.		Weight, in running order. POUNDS.			Separate Tender.
	INCHES.	INCHES.	Of Driving Wheels.	Total.	On Driving Wheels. LBS.	On Truck. LBS.	Total. LBS.	Capacity of Tank. GALS.
1	19 x 24	62	13 ft. 6 in.	24 ft. 6 in.	92,000	32,000	124,000	3,500
2	19 x 26	54	12 ft. 6 in.	23 ft. 0 in.	100,000	30,000	130,000	3,500
3								
4								
5								

Load in tons of 2,000 pounds in addition to Engine and Tender, at 35 miles an hour, on a grade per mile of

	On a Level.	10 ft.	20 ft.	40 ft.	60 ft.	80 ft.	100 ft.	125 ft.	150 ft.
1	1,620	1,137	869	572	415	317	252	194	151
2	2,530	1,720	1,284	835	607	468	376	300	238
3									
4									
5									

PLATE XXIII.

Ten Wheel Locomotives

FOR PASSENGERS.

Gauge, 4 ft., 8½ in., or wider. Fuel, Bituminous Coal.

General Design shown by Plate XXIII.

	Cylinders. Diameter and Stroke. INCHES.	Dia'eter of Driving Wheels. INCHES.	Wheel Base. Of Driving Wheels.	Wheel Base. Total.	Weight, in running order. POUNDS. On Driving Wheels. LBS.	Weight, in running order. POUNDS. On Truck. LBS.	Weight, in running order. POUNDS. Total. LBS.	Separate Tender. Capacity of Tank. GALS.
1	19 x 24	66	13 ft. 0 in.	23 ft. 9 in.	96,000	30,000	126,000	4,000
2	19 x 24	72	13 ft. 0 in.	23 ft. 9 in.	97,000	30,000	127,000	4,000
3	20 x 24	63	13 ft. 0 in.	23 ft. 9 in.	100,000	30,000	130,000	4,000
4								
5								

Load in tons of 2,000 pounds in addition to Engine and Tender, at 45 miles an hour, on a grade per mile of

	On a Level.	10 ft.	20 ft.	40 ft.	60 ft.	80 ft.	100 ft.	125 ft.	150 ft.
1	704	629	520	375	288	235	179	135	103
2	632	567	468	337	256	207	156	114	85
3	826	744	618	443	350	287	218	167	129
4									
5									

PLATE XXIV.

Ten Wheel Locomotives

FOR PASSENGERS.

Gauge, 4 ft. 8½ in., or wider. Fuel, Bituminous Coal.

General Design shown by Plate XXIV.

	Cylinders. Diameter and Stroke.	Diam'ter of Driving Wheels.	Wheel Base.		Weight, in running order. POUNDS.			Separate Tender.
			Of Driving Wheels.	Total.	On Driving Wheels.	On Truck.	Total.	Capacity of Tank.
	INCHES.	INCHES.			LBS.	LBS.	LBS.	GALS.
1	19½ x 26	69	13 ft. 3 in.	24 ft. 3 in.	109,000	33,000	142,000	4,200
2								
3								
4								
5								

Load in tons of 2,000 pounds in addition to Engine and Tender, at 30 miles an hour, on a grade per mile of

	On a Level.	10 ft.	20 ft.	40 ft.	60 ft.	80 ft.	100 ft.	125 ft.	150 ft.
1	1,000	705	570	400	301	237	179	135	103
2									
3									
4									
5									

PLATE XXV.

Ten Wheel Locomotives

FOR PASSENGERS.

Gauge, 4 ft., 8½ in. or wider. Fuel, Bituminous or Anthracite Coal.

General Design shown by Plate XXV.

	Cylinders Diameter and Stroke.	Dia'eter of Driving Wheels.	Wheel Base.		Weight, in running order. POUNDS.			Separate Tender.
			Of Driving Wheels.	Total.	On Driving Wheels.	On Truck.	Total.	Capacity of Tank.
	INCHES.	INCHES.			LBS.	LBS.	LBS.	GALS.
1	20 x 24	72	13 ft. 6 in.	24 ft. 9 in.	110,000	40,000	150,000	4,000
2								
3								
4								
5								

	Load in tons of 2,000 pounds in addition to Engine and Tender, at 50 miles an hour, on a grade per mile of								
	On a Level.	10 ft.	20 ft.	40 ft.	60 ft.	80 ft.	100 ft.	125 ft.	150 ft.
1	829	646	502	362	270	213	156	120	93
2									
3									
4									
5									

PLATE XXVI.

Ten Wheel Locomotives

FOR PASSENGERS.

Gauge, 4 ft., 8½ in., or wider. Fuel, Bituminous Coal.

General Design shown by Plate XXVI.

	Cylinders. Diameter and Stroke. INCHES.	Dia'eter of Driving Wheels. INCHES.	Wheel Base. Of Driving Wheels.	Wheel Base. Total.	Weight, in running order. POUNDS. On Driving Wheels. LBS.	On Truck. LBS.	Total. LBS.	Separate Tender. Capacity of Tank. GALS.
1	19½ x 26	69	13 ft. 3 in.	24 ft. 3 in.	109,000	35,500	144,500	4,200
2								
3								
4								
5								

Load in tons of 2,000 pounds in addition to Engine and Tender, at 50 miles an hour, on a grade per mile of

	On a Level.	10 ft.	20 ft.	40 ft.	60 ft.	80 ft.	100 ft.	125 ft.	150 ft.
1	1,000	705	570	400	301	237	179	135	103
2									
3									
4									
5									

PLATE XXVII.

Consolidation Locomotive

FOR FREIGHT.

Gauge, 4 ft., 8½ in., or wider Fuel, Bituminous Coal.

General Design shown by Plate XXVII.

	Cylinders. Diameter and Stroke.	Diam'ter of Driving Wheels.	Wheel Base.		Weight, in running order. POUNDS.			Separate Tender.
			Of Driving Wheels.	Total.	On Driving Wheels.	On Truck.	Total.	Capacity of Tank.
	INCHES.	INCHES.			LBS.	LBS.	LBS.	GALS.
1	21 x 24	56	16 ft. 9 in.	24 ft. 5 in.	119,000	18,000	137,000	3,800
2	21 x 26	54	15 ft. 2 in.	24 ft. 3 in.	133,000	16,000	149,000	4,000
3	22 x 26	60	16 ft. 9 in.	24 ft. 9 in.	144,000	16,000	160,000	4,000
4								
5								

Load in tons of 2,000 pounds in addition to Engine and Tender, at 15 miles an hour, on a grade per mile of

	On a Level.	10 ft.	20 ft.	40 ft.	60 ft.	80 ft.	100 ft.	125 ft.	150 ft.
1	2,916	2,366	1,706	1,107	799	616	496	370	318
2	3,270	2,655	1,915	1,242	897	690	558	456	357
3	3,235	2,625	1,900	1,226	886	683	549	440	351
4									
5									

PLATE XXVIII.

Consolidation Locomotive

FOR FREIGHT.

Gauge, 4 ft., 8½ in., or wider. Fuel, Bituminous Coal.

General Design shown by Plate XXVIII.

	Cylinders. Diameter and Stroke. INCHES.	Diameter of Driving Wheels. INCHES.	Wheel Base. Of Driving Wheels.	Wheel Base. Total.	Weight, in running order. POUNDS. On Driving Wheels. LBS.	On Truck. LBS.	Total. LBS.	Separate Tender. Capacity of Tank. GALS.
1	20 x 24	54	15 ft. 2 in.	22 ft. 10 in.	117,000	16,000	133,000	4,000
2								
3								

Load in tons of 2,000 pounds in addition to Engine and Tender, at 15 miles an hour, on a grade per mile of

	On a Level.	10 ft.	20 ft.	40 ft.	60 ft.	80 ft.	100 ft.	125 ft.	150 ft.
1	2,710	2,222	1,624	1,037	748	575	464	369	295
2									
3									

PLATE XXIX

Consolidation Locomotive

FOR FREIGHT.

Gauge, 4 ft., 8½ in., or wider. Fuel, Bituminous Coal.

General Design shown by Plate XXIX.

	Cylinders. Diameter and Stroke. INCHES.	Diam'ter of Driving Wheels. INCHES.	Wheel Base. Of Driving Wheels.	Wheel Base. Total.	Weight, in running order. POUNDS. On Driving Wheels. LBS.	Weight, in running order. POUNDS. On Truck. LBS.	Weight, in running order. POUNDS. Total. LBS.	Separate Tender. Capacity of Tank. GALS.
1	20 x 26	58	15 ft. 10 in.	23 ft. 6 in.	130,700	14,300	145,000	4,000
2	21 x 24	51	14 ft. 8 in.	22 ft. 5 in.	128,000	16,000	144,000	4,000
3	22 x 26	56	15 ft. 10 in.	23 ft. 6 in.	142,000	20,000	160,000	4,200
4								
5								

Load in tons of 2,000 pounds in addition to Engine and Tender, at 20 miles an hour, on a grade per mile of

	On a Level.	10 ft.	20 ft.	40 ft.	60 ft.	80 ft.	100 ft.	125 ft.	150 ft.
1	2,620	1,776	1,323	860	620	480	385	310	240
2	2,950	2,010	1,500	980	715	554	450	375	290
3	3,600	2,425	1,910	1,208	895	684	544	445	350
4									
5									

PLATE XXV.

Decapod Locomotive

FOR FREIGHT.

Gauge, 4 ft., 8½ in., or wider. Fuel, Bituminous Coal.

General Design shown by Plate XXX.

	Cylinders. Diameter and Stroke. INCHES.	Dia'eter of Driving Wheels. INCHES.	Wheel Base. Of Driving Wheels.	Wheel Base. Total.	Weight, in running order. POUNDS. On Driving Wheels. LBS.	Weight, in running order. On Track. LBS.	Weight, in running order. Total. LBS.	Separate Tender. Capacity of Tank. GALS.
1	22 x 28	50	17 ft. 10 in.	17 ft. 10 in.	150,000	—	150,000	4,000
2	22 x 28	56	19 ft. 10 in.	19 ft. 10 in.	156,000	—	156,000	4,000
3								
4								
5								

Load in tons of 2,000 pounds in addition to Engine and Tender, at 15 miles an hour, on a grade per mile of

	On a Level.	10 ft.	20 ft.	40 ft.	60 ft.	80 ft.	100 ft.	125 ft.	150 ft.
1	3,950	3,212	2,358	1,519	1,105	859	697	563	457
2	3,687	2,913	2,135	1,370	993	767	621	499	403
3									
4									
5									

PLATE XXXI

Four Wheel Locomotives

FOR SWITCHING.

Gauge, 4 ft., 8½ in., or wider. Fuel, Bituminous or Anthracite Coal.

General Design shown by Plate XXXI.

	Cylinders. Diameter and Stroke. INCHES.	Dia'eter of Driving Wheels. INCHES.	Wheel Base. Of Driving Wheels.	Wheel Base. Total.	Weight, in running order. POUNDS. On Driving Wheels. LBS.	On Truck. LBS.	Total. LBS.	Separate Tender. Capacity of Tank. GALS.
0	15 x 24	48	7 ft. 6 in.	7 ft. 6 in.	68,000	—	68,000	1,800
1	16 x 24	50	8 ft. 0 in.	8 ft. 0 in.	75,000	—	75,000	2,000
2	17 x 24	50	8 ft. 0 in.	8 ft. 0 in.	79,000	—	79,000	2,200
3								
4								

Load in tons of 2,000 pounds in addition to Engine and Tender, at 10 miles an hour, on a grade per mile of

	On a Level.	10 ft.	20 ft.	40 ft.	60 ft.	80 ft.	100 ft.	125 ft.	150 ft.
0	2,661	1,648	1,176	736	547	405	337	260	211
1	2,913	1,800	1,287	808	600	447	361	288	235
2	3,292	2,039	1,455	913	656	505	410	326	267
3									
4									

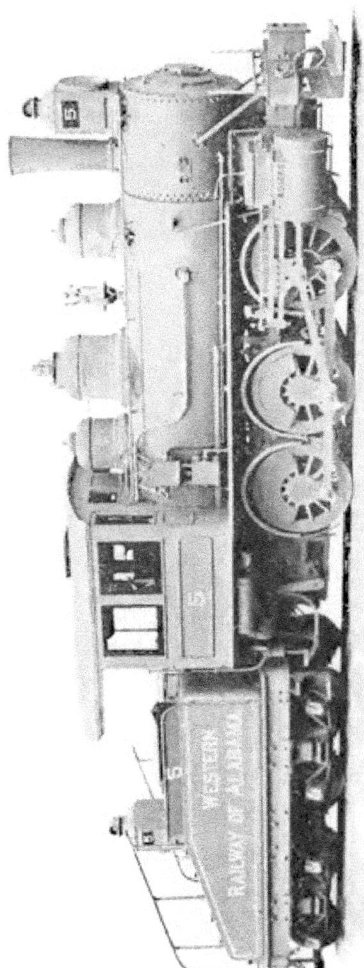

PLATE XXXII.

Six Wheel Locomotives

FOR SWITCHING.

Gauge, 4 ft., 8½ in., or wider. Fuel, Bituminous or Anthracite Coal

General Design shown by Plate XXXII.

	Cylinders. Diameter and Stroke. INCHES.	Dia'cter of Driving Wheels INCHES.	Wheel Base. Of Driving Wheels.	Wheel Base. Total.	Weight, in running order. POUNDS. On Driving Wheels. LBS.	On Truck. LBS.	Total. LBS.	Separate Tender. Capacity of Tank. GALS.
1	16 x 24	50	9 ft. 6 in.	9 ft. 6 in.	79,000	—	79,000	2,000
2	17 x 24	50	10 ft. 0 in.	10 ft. 0 in.	85,000	—	85,000	2,200
3	18 x 24	50	10 ft. 6 in.	10 ft. 6 in.	93,000	—	93,000	2,200
4								
5								

Load in tons of 2,000 pounds in addition to Engine and Tender, at 10 miles an hour, on a grade per mile of

	On a Level.	10 ft.	20 ft.	40 ft.	60 ft.	80 ft.	100 ft.	125 ft.	150 ft.
1	2,911	1,798	1,285	806	598	445	359	286	233
2	3,288	2,035	1,451	909	652	501	405	322	262
3	3,671	2,273	1,588	1,018	725	564	453	363	296
4									
5									

PLATE XXXVIII.

Four Wheel Tank Locomotives

FOR SWITCHING.

Gauge, 4 ft., 8½ in., or wider   Fuel, Bituminous Coal.

General Design shown by Plate XXXIII.

	Cylinders. Diameter and Stroke.	Diam'ter of Driving Wheels.	Wheel Base.		Weight, in running order. POUNDS.			Separate Tender.
			Of Driving Wheels.	Total.	On Driving Wheels.	On Truck.	Total.	Capacity of Tank.
	INCHES.	INCHES.			LBS.	LBS.	LBS.	GALS.
1	12 x 22	48	7 ft. 6 in.	7 ft. 6 in.	60,000	—	60,000	500
2	14 x 24	50	8 ft. 0 in.	8 ft. 0 in.	74,000	—	74,000	600
3								

Load in tons of 2,000 pounds in addition to Engine and Tender, at 10 miles an hour, on a grade per mile of

	On a Level.	10 ft.	20 ft.	40 ft.	60 ft.	80 ft.	100 ft.	125 ft.	150 ft.
1	1,112	902	650	412	298	230	189	152	124
2	1,575	1,279	911	571	411	318	257	206	169
3									

PLATE XXXVI.

Six Wheel Tank Locomotives

FOR SWITCHING.

Gauge, 4 ft., 8½ in., or wider. Fuel, Bituminous Coal.

General Design shown by Plate XXXIV.

	Cylinders. Diameter and Stroke. INCHES.	Diam'ter of Driving Wheels. INCHES.	Wheel Base. Of Driving Wheels.	Wheel Base. Total.	Weight, in running order. POUNDS. On Driving Wheels. LBS.	Weight, in running order. On Truck. LBS.	Weight, in running order. Total. LBS.	Separate Tender. Capacity of Tank. GALS.
1	12 x 18	39	10 ft. 0 in.	10 ft. 0 in.	60,000	—	60,000	500
2	15 x 18	39	10 ft. 0 in.	10 ft. 0 in.	81,000	—	81,000	700
3	15 x 22	44	10 ft. 0 in.	10 ft. 0 in.	86,000	—	86,000	750
4	16 x 22	46	10 ft. 0 in.	10 ft. 0 in.	90,000	—	90,000	800
5	17 x 24	50	10 ft. 6 in.	10 ft. 6 in.	96,000	—	96,000	850

Load in tons of 2,000 pounds in addition to Engine and Tender, at 10 miles an hour, on a grade per mile of

	On a Level.	10 ft.	20 ft.	40 ft.	60 ft.	80 ft.	100 ft.	125 ft.	150 ft.
1	1,034	840	605	383	275	212	172	135	112
2	1,630	1,320	958	560	438	340	275	219	181
3	1,630	1,320	958	560	438	340	275	220	182
4	1,916	1,555	1,123	711	517	401	326	260	215
5	2,179	1,767	1,278	812	589	457	371	296	245

PLATE XXXI.

Six Wheel Tank Locomotives

(FOUR DRIVING AND TWO TRAILING WHEELS.)

FOR SWITCHING.

ON CURVES AS SHARP AS 60 FEET RADIUS.

Gauge, 4 ft., 8½ in., or wider. Fuel, Bituminous Coal.

General Design shown by Plate XXXV.

	Cylinders. Diameter and Stroke. INCHES.	Diameter of Driving Wheels. INCHES.	Wheel Base. Of Driving Wheels.	Wheel Base. Total.	Weight in running order. POUNDS. On Driving Wheels. LBS.	Weight in running order. On Track. LBS.	Weight in running order. Total. LBS.	Tank on Engine. Capacity of Tank. GALS.
1	16 x 24	44	6 ft. 0 in.	13 ft. 6 in.	74,000	11,000	85,000	500

Load in tons of 2,000 pounds in addition to Engine and Tender, at 10 miles an hour, on a grade per mile of

	On a Level	10 ft.	20 ft.	40 ft.	60 ft.	80 ft.	100 ft.	125 ft.	150 ft.
1	1,850	1,368	1,068	722	547	426	358	287	242

PLATE XXXVI.

Eight Wheel Double-Ender

WITH TANK OVER REAR TRUCK.

FOR SUBURBAN PASSENGER SERVICE.

Gauge, 4 ft., 8½ in., or wider. Fuel, Bituminous Coal.

General Design shown by Plate XXXVI.

	Cylinders. Diameter and Stroke. INCHES.	Diameter of Driving Wheels. INCHES.	Wheel Base. On Driving Wheels.	Wheel Base. Total.	Weight, in running order. POUNDS. On Driving Wheels. LBS.	On Front and Rear Truck. LBS.	Total. LBS.	Separate Tender. Capacity of Tank. GALS.
1	13 x 22	44	5 ft. 6 in.	18 ft. 6 in.	45,000	{ 10,000 { 17,000	72,000	700

Load in tons of 2,000 pounds in addition to Engine and Tender, at 25 miles an hour, on a grade per mile of

	On a Level.	10 ft.	20 ft.	40 ft.	60 ft.	80 ft.	100 ft.	125 ft.	150 ft.
1	900	700	550	375	275	210	180	140	110

PLATE XXXVII.

10 Wheel Double-Ender Locomotive

WITH TANK OVER REAR TRUCK.

Gauge, 4 ft., 8½ in., or wider. Fuel, Bituminous Coal.

General Design shown by Plate XXXVII.

	Cylinders. Diameter and Stroke. INCHES.	Dia'eter of Driving Wheels. INCHES.	Wheel Base. Of Driving Wheels.	Wheel Base. Total.	Weight, in running order. POUNDS. On Driving Wheels. LBS.	Weight, in running order. POUNDS. On Front and Rear Truck LBS.	Weight, in running order. POUNDS. Total. LBS.	Tank on Engine. Capacity of Tank. GALS.
1	15 x 22	44	9 ft. 9 in.	21 ft. 8 in.	63,000	{ 12,500 { 12,500	88,000	700

Load in tons of 2,000 pounds in addition to Engine and Tender, at 25 miles an hour, on a grade per mile of

	On a Level.	10 ft.	20 ft.	40 ft.	60 ft.	80 ft.	100 ft.	125 ft.	150 ft.
1	1,100	810	630	[illegible]	314	242	200	156	130

PLATE XXXVIII.

Eight Wheel Forney Engine

WITH TANK OVER TRUCK.

Gauge, 4 ft., 8½ in., or wider. Fuel, Anthracite or Bituminous Coal.

General Design shown by Plate XXXVIII.

	Cylinders. Diameter and Stroke. INCHES.	Dia'eter of Driving Wheels. INCHES.	Wheel Base. Of Driving Wheels.	Wheel Base. Total.	Weight, in running order. POUNDS. On Driving Wheels. LBS.	On Truck. LBS.	Total. LBS.	Tank on Engine. Capacity of Tank. GALS.
1	11 x 16	42	5 ft. 0 in.	16 ft. 1 in.	30,000	14,000	44,000	500
2	12 x 18	42	5 ft. 3 in.	16 ft. 7 in.	36,000	18,000	54,000	650
3								
4								
5								

Load in tons of 2,000 pounds in addition to Engine and Tender, at 25 miles an hour, on a grade per mile of

	On a Level.	10 ft.	20 ft.	40 ft.	60 ft.	80 ft.	100 ft.	125 ft.	150 ft.
1	478	340	255	168	124	96	75	60	50
2	600	456	342	227	170	130	106	83	68
3									
4									
5									

PLATE XXXIX.

12 Wheel Double-Ender Locomotives

WITH TANK OVER REAR TRUCK.

FOR SUBURBAN PASSENGER SERVICE.

Gauge, 4 ft., 8½ in., or wider. Fuel, Bituminous or Anthracite Coal.

General Design shown by Plate XXXIX.

	Cylinders. Diameter and Stroke. INCHES.	Diam'ter of Driving Wheels. INCHES.	Wheel Base. Of Driving Wheels.	Wheel Base. Total.	Weight, in running order. POUNDS. On Driving Wheels. LBS.	Weight, in running order. POUNDS. On Front and Rear Truck. LBS.	Weight, in running order. POUNDS. Total. LBS.	Tank on Engine. Capacity of Tank. GALS.
1	14 x 22	54	6 ft. 3 in.	27 ft. 0 in.	50,000	13,000 36,000	99,000	1,000
2	15 x 22	56	6 ft. 6 in.	27 ft. 6 in.	60,000	14,500 40,000	114,500	1,500
3	16 x 22	56	7 ft. 0 in.	28 ft. 0 in.	67,000	16,000 45,000	128,000	1,800
4	17 x 24	56	6 ft. 10 in.	32 ft. 7 in.	68,000	19,000 58,000	145,000	2,500
5								

Load in tons of 2,000 pounds in addition to Engine and Tender, at 30 miles an hour, on a grade per mile of

	On a Level.	10 ft.	20 ft.	40 ft.	60 ft.	80 ft.	100 ft.	125 ft.	150 ft.
1	730	526	412	282	204	166	126	98	77
2	796	580	454	310	224	182	138	107	85
3	900	660	516	348	255	207	158	120	95
4	1,100	818	612	435	320	262	200	153	123
5									

PLATE XL.

12 Wheel Double-Ender Locomotives

WITH TANK OVER REAR TRUCK.

FOR SUBURBAN PASSENGER SERVICE.

Gauge, 4 ft., 8½ in., or wider. Fuel, Bituminous or Anthracite Coal.

General Design shown by Plate XI.

	Cylinders. Diameter and Stroke.	Diam'ter of Driving Wheels.	Wheel Base.		Weight, in running order. POUNDS.			Separate Tender.
			Of Driving Wheels.	Total.	On Driving Wheels.	On Truck.	Total.	Capacity of Tank.
	INCHES.	INCHES.			LBS.	LBS.	LBS.	GALS.
1	17 x 24	56	13 ft. 0 in.	33 ft. 5 in.	82,000	16,000 46,000	144,000	2,000
2	18 x 24	56	13 ft. 0 in.	33 ft. 9 in.	92,000	18,000 50,000	160,000	2,400
3								
4								
5								

Load in tons of 2,000 pounds in addition to Engine and Tender, at 30 miles an hour, on a grade per mile of

	On a Level.	10 ft.	20 ft.	40 ft.	60 ft.	80 ft.	100 ft.	125 ft.	150 ft.
1	1,100	818	642	435	320	262	200	153	124
2	1,280	920	722	490	360	290	226	174	140
3									
4									
5									

PLATE XLI.

Eight Wheel Standard Locomotives

FOR PASSENGERS OR FREIGHT.

Narrow Gauge Track. Fuel, Bituminous Coal or Wood.

General Design shown by Plate XLI.

	Cylinders. Diameter and Stroke.	Dia'eter of Driving Wheels.	Wheel Base.		Weight, in running order. POUNDS.			Separate Tender.
			Of Driving Wheels.	Total.	On Driving Wheels.	On Truck.	Total.	Capacity of Tank.
	INCHES.	INCHES.			LBS.	LBS.	LBS.	GALS.
1	12 x 18	39	6 ft. 2 in.	17 ft. 4 in.	31,400	17,100	48,500	1,500
2	14 x 20	54	7 ft. 6 in.	20 ft. 5 in.	38,000	22,000	60,000	1,800
3								
4								
5								

	Load in tons of 2,000 pounds in addition to Engine and Tender, at 20 miles an hour, on a grade per mile of								
	On a Level.	10 ft.	20 ft.	40 ft.	60 ft.	80 ft.	100 ft.	125 ft.	150 ft.
1	952	769	550	332	242	183	145	112	88
2	1,028	827	588	361	252	187	146	109	84
3									
4									
5									

PLATE XLII.

Special Locomotives

(FOUR DRIVING AND TWO TRAILING WHEELS.)

FOR FREIGHT.

Narrow Gauge Track. Fuel, Bituminous or Anthracite Coal.

General Design shown by Plate XLII.

	Cylinders. Diameter and Stroke. INCHES.	Dia'eter of Driving Wheels. INCHES.	Wheel Base. Of Driving Wheels.	Wheel Base. Total.	Weight, in running order. POUNDS. On Driving Wheels. LBS.	Weight, in running order. POUNDS. On Truck. LBS.	Weight, in running order. POUNDS. Total. LBS.	Separate Tender. Capacity of Tank. GALS.
1	12 x 18	36	6 ft. 0 in.	13 ft. 8 in.	46,000	6,000	52,000	1,500
2	14 x 20	36	6 ft. 0 in.	13 ft. 8 in.	56,000	8,000	64,000	2,000
3								
4								
5								

Load in tons of 2,000 pounds in addition to Engine and Tender, at 10 miles an hour, on a grade per mile of

	On a Level.	10 ft.	20 ft.	40 ft.	60 ft.	80 ft.	100 ft.	125 ft.	150 ft.
1	1,110	897	643	402	286	218	173	135	108
2	1,694	1,371	986	621	446	341	275	216	175
3									
4									
5									

PLATE XLIII.

Mogul Locomotives

FOR FREIGHT.

Narrow Gauge Track. Fuel, Bituminous Coal or Wood.

General Design shown by Plate XLIII.

	Cylinders. Diameter and Stroke.	Dia'eter of Driving Wheels.	Wheel Base.		Weight, in running order. POUNDS.			Separate Tender.
			Of Driving Wheels.	Total.	On Driving Wheels.	On Truck.	Total.	Capacity of Tank.
	INCHES.	INCHES.			LBS.	LBS.	LBS.	GALS.
1	14 x 18	39	14 ft. 6 in.	20 ft. 9 in.	51,000	12,000	63,000	2,000
2	15 x 18	39	14 ft. 6 in.	20 ft. 9 in.	56,000	11,000	70,000	2,200
3	17 x 22	48	11 ft. 6 in.	18 ft. 8 in.	76,000	9,500	85,500	2,600
4								
5								

Load in tons of 2,000 pounds in addition to Engine and Tender, at 15 miles an hour, on a grade per mile of

	On a Level.	10 ft.	20 ft.	40 ft.	60 ft.	80 ft.	100 ft.	125 ft.	150 ft.
1	1,528	997	728	464	334	256	205	161	130
2	1,862	1,219	994	574	416	322	260	207	169
3	1,660	1,360	992	631	452	346	276	219	172
4									
5									

PLATE XLIV.

Mogul Locomotives

FOR FREIGHT.

Narrow Gauge Track. Fuel, Bituminous Coal or Wood.

General Design shown by Plate XLIV.

	Cylinders. Diameter and Stroke. INCHES.	Dia'eter of Driving Wheels. INCHES.	Wheel Base. Of Driving Wheels.	Wheel Base. Total.	Weight, in running order. POUNDS. On Driving Wheels. LBS.	On Truck. LBS.	Total. LBS.	Separate Tender. Capacity of Tank. GALS.
1	12 x 18	36	8 ft. 9 in.	14 ft. 6 in.	36,000	8,000	44,000	1,500
2	13 x 18	36	8 ft. 9 in.	14 ft. 6 in.	40,000	10,000	50,000	1,800
3								
4								
5								

Load in tons of 2,000 pounds in addition to Engine and Tender, at 15 miles an hour, on a grade per mile of

	On a Level.	10 ft.	20 ft.	40 ft.	60 ft.	80 ft.	100 ft.	125 ft.	150 ft.
1	1,020	662	480	302	215	163	102	100	78
2	1,200	780	566	357	254	193	153	116	92
3									
4									
5									

PLATE XLV.

Four Wheel Tank Locomotive

FOR SWITCHING.

Narrow Gauge Track. Fuel, Bituminous Coal or Wood.

General Design shown by Plate XLV.

	Cylinders. Diameter and Stroke	Diam'ter of Driving Wheels.	Wheel Base.		Weight, in running order. POUNDS.			Separate Tender.
	INCHES.	INCHES.	Of Driving Wheels.	Total.	On Driving Wheels. LBS.	On Truck. LBS.	Total. LBS.	Capacity of Tank. GALS.
1	8 x 12	26	5 ft.	5 ft.	19,000	—	19,000	200
2	8 x 12	30	5 ft.	5 ft.	19,500	—	19,500	200
3								
4								
5								

Load in tons of 2,000 pounds in addition to Engine and Tender, at 6 miles an hour, on a grade per mile of

	On a Level.	10 ft.	20 ft.	40 ft.	60 ft.	80 ft.	100 ft.	125 ft.	150 ft.
1	703	434	310	195	141	109	89	72	59
2	608	375	267	168	121	94	76	61	50
3									
4									
5									

www.ingramcontent.com/pod-product-compliance
Lightning Source LLC
Chambersburg PA
CBHW021938190326
41519CB00009B/1060